Law and the Dead

The governance of the dead in the eighteenth and nineteenth centuries gave rise to a new arrangement of thanato-politics in the West. Legal, medical and bureaucratic institutions developed innovative technologies for managing the dead, maximising their efficacy and exploiting their vitality. *Law and the Dead* writes a history of their institutional life in the nineteenth and twentieth centuries.

With a particular focus on the technologies of the death investigation process, including place-making, the forensic gaze, bureaucratic manuals, record-keeping and radiography, this book examines how the dead came to be incorporated into legal institutions in the modern era. Drawing on the writings of philosophers, historians and legal theorists, it offers tools for thinking through how the dead dwell in law, how their lives persist through the conduct of office, and how coroners assume responsibility for taking care of the dead.

This historical and interdisciplinary book offers a provocative challenge to conventional thinking about the sequestration of the dead in the nineteenth and twentieth centuries. It asks the reader to think through and with legal institutions when writing a history of the dead, and to trace the important role assumed by coroners in the governance of the dead. This book will be of interest to scholars working in law, history, sociology and criminology.

Dr Marc Trabsky is a Senior Lecturer in Law at La Trobe University, Melbourne.

Law and the Dead
Technology, Relations and Institutions

Marc Trabsky

Routledge
Taylor & Francis Group

a GlassHouse book

First published 2019
by Routledge
2 Park Square, Milton Park, Abingdon, Oxon OX14 4RN

and by Routledge
52 Vanderbilt Avenue, New York, NY 10017

First issued in paperback 2020

A GlassHouse book

Routledge is an imprint of the Taylor & Francis Group, an informa business

British Library Cataloguing-in-Publication Data
A catalogue record for this book is available from the British Library

Library of Congress Cataloging-in-Publication Data
A catalog record has been requested for this book

ISBN 13: 978-0-367-66024-6 (pbk)
ISBN 13: 978-0-8153-7523-4 (hbk)

Typeset in Galliard
by Swales & Willis, Exeter, Devon, UK

For Jacinthe Flore

Contents

Acknowledgements

The development of this book has spanned a great number of years and I am indebted to the many individuals who provided intellectual generosity, critical support and challenging discourse. The book began life as a doctoral thesis at the University of Melbourne, which was supported by an Australian Government Research Training Program Scholarship and a Melbourne Research Scholarship. I am appreciative of the mentorship provided by Shaun McVeigh and Peter Rush, and the collegiality of a diverse group of scholars: Tom Andrews, Olivia Barr, Kathleen Birrell, Sara Dehm, Ann Genovese, Jake Goldenfein, Vicki Huang, Dave McDonald, Yoriko Otomo, James Parker, Laura Petersen, Megan Richardson, Amanda Scardamaglia and Cait Storr.

Laura Griffin has not only been a passionate reader and astute editor, but also a provocative interlocutor. For creating a dynamic intellectual community, where interdisciplinary research is enthusiastically pursued, I thank my colleagues at La Trobe University: Paula Baron, Tobias Barkley, Madelaine Chiam, Pascale Chifflet, Kirsty Duncanson, Julia Dehm, Maria Elander, Laura Griffin, Emma Henderson, Fiona Kelly, Patrick Keyzer, Anita Mackay, Jill Murray, Tarryn Phillips, Hannah Robert, Savitri Taylor, Raul Sanchez Urribarri and David Wishart.

The book has benefited from numerous conversations, too many to properly acknowledge here, shared at the margins of conferences, workshops and symposiums since 2012. It has most recently prospered from seminars held at Flinders University, Griffith University, Manchester University, the University of Kent and University of Newcastle. For their insightful comments and engaging discourse, I thank Margot Brazier, Ian Burney, Margaret Davies, Maria Drakopoulou, John Flood, Maria Giannacopoulos, Danielle Griffith, Hyo Yoon Kang, Ed Kirton-Darling, Angela Melville, Rose Parfitt, Nick Piska and Kevin Sobel-Read. Connal Parsley's hospitality during my visit at Kent Law School, and his incisive remarks in relation to Chapter 2, were invaluable. I am grateful for my long-standing friendship with Edward Mussawir, who has inspired me to change course at critical junctures, and Rebecca Scott Bray's unwavering enthusiasm and promotion of my scholarship in coronial studies.

I wish to thank the staff at the Public Records Office of Victoria, Victorian Police Museum, Victorian Institute of Forensic Medicine Library, State Library of Victoria, Special Collections at the Baillieu Library, the University of

Melbourne and the Wellcome Library in the United Kingdom for providing valuable assistance for accessing their archives. In the final year of writing, the book was generously funded by a La Trobe University Transforming Human Societies grant and a La Trobe University Social Research Assistance Platform grant. Meredith Jones provided exceptional research assistance for Chapter 5 and Julia Farrell expertly copyedited and indexed the final manuscript. I am indebted to the assistance provided by Routledge, particularly the encouragement of Colin Perrin, and Nicola Sharpe for finalising the manuscript for publication.

My family and friends have given me solace during this long journey. Ohad Kozminsky, Bella Li and Agata Wierzbowski have challenged the way I think about art, philosophy and law. My parents, Karen and Morris, and my sister Dani, have supported me through the turbulence of my scholarly pursuits. Jacinthe Flore has been my most zealous interlocutor throughout the entire length of the project. She has been patient, critical and caring. I couldn't have completed this enormous task without her and I cannot capture in words the incredible support that she has given me.

Parts of this book have been published elsewhere, and I am thankful for the permission to include those publications here: an early version of Chapter 1 in 'Walking with the Dead: Coronial Law and Spatial Justice in the Necropolis', in Edward Mussawir and Chris Butler (eds), *Spaces of Justice: Positions, Passages, Appropriations* (Routledge, 2017); a portion of Chapter 2 was taken from 'The Custodian of Memories: Coronial Architecture in Nineteenth Century Melbourne' (2015) 24(2) *Griffith Law Review* 199; and an early version of Chapter 3 from 'The Coronial Manual and the Bureaucratic Logic of the Coroner's Office' (2016) 12(2) *International Journal of Law in Context* 195.

Introduction

'... one does not get rid of the dead, one is never finished with them'.[1]

This book writes a history of the institutional life of law and the dead in the nineteenth and twentieth centuries. With a particular focus on the death investigation process, it examines how the dead became incorporated into a type of legal institution, that being the office of coroner, during a period when it underwent significant transformations. The office of coroner has occupied an important role in the common law since the twelfth century. Its status may have waned and its duties may have changed over time, yet its enduring concern with investigating sudden, unnatural or violent death has preserved its vital role in the governance of the dead. *Law and the Dead* offers a historical account of the modalities by which coroners have formed legal relations with the dead. It does so by examining law in terms of its technologies and its institutional formations. The chapters that follow present an institutional history of the life of law by thinking through how technologies have attached the dead to legal institutions and how legal officers have assumed responsibilities for caring for the dead.

The governance of the dead

Many events precipitated the abolition of the deodand in the nineteenth century. The most notable was the Sonning Cutting railway accident of 1841, which had a profound effect on the political appetite for legislative reform. On 24 December 1841, a goods train travelling from London Paddington to Bristol Temple Mead derailed near Reading, resulting in nine deaths and 16 injured passengers. The coronial inquest that immediately followed the accident found that a bank slip in the cutting led to the derailment, while the deaths were caused by the proximity of the coaches to the engine. The jury, much to the chagrin of the coroner, held the owner of the train responsible for causing the

---

1 Paul Ricoeur, *Living Up To Death* (David Pellauer trans, The University of Chicago Press, 2009) 9.

derailment and declared that the train, its engine, tender and carriage constituted a deodand, which they appraised at £1,000. They required Great Western Railways to pay the value of the deodand to the Crown, who would presumably forward it as compensation to the families of the deceased.[2] However, the company successfully appealed against the inquest on a technicality, resulting in public outcry and media furore about the difficulties of holding train owners accountable for railway deaths. The Sonning Cutting accident inflamed political debates about the equity, utility and economics of the deodand in the age of mechanised transportation.

The law of deodands posed a fiscal problem for the railway industry. It prescribed that any animate or inanimate object that moved towards the death of a person was to be forfeited to the sovereign: '[t]hus a mill-wheel was often deodand but not the whole mill, a ship but not its cargo, or a branch and not the whole tree'.[3] Certainly a train, but not its goods, would fit this vague definition. With the rise of urbanisation and the invention of steam power and machine-based production, the number of fatal accidents caused by movable entities increased exponentially in the first half of the nineteenth century. Without any recourse to financial compensation, families of the deceased resorted to the remedy of the deodand, hoping that coronial jurors would hold capitalists responsible for the deaths their chattels caused and compel them to properly recompense families for the loss of the deceased. During the period of industrialisation in Britain, as labourers eschewed horses for trains and moved from rural farms to urban factories, the deodand threatened to financially cripple railway owners. It is no wonder then that they strenuously lobbied Parliament to put an end to what was believed to be an anachronistic law. The invention of the railway accident preceded both the *Deodands Act 1846* (UK) and the *Fatal Accidents Act 1846* (UK), which abolished the law of deodands and replaced it with a right for bereaved relatives to claim damages following the death of the plaintiff.[4]

2 Legal historians point to the origins of the deodand in the twelfth century, though its derivation from the Latin *deo donatus* (a gift to God) suggests that it could have appeared much earlier. Its initial purpose was to raise revenue for the King; however, over time, the Crown passed the value of the chattel to the deceased's relatives or a charitable institution. For a critique of the contingency of the deodand as a remedy for families of the deceased, see Edward Kirton-Darling, 'Searching for Pigeons in the Belfry: The Inquest, the Abolition of the Deodand and the Rise of the Family' (2014) *Law, Culture and the Humanities* 1, 8–9. http://journals.sagepub.com/doi/full/10.1177/1743872114560701.

3 R.F. Hunnisett, *The Medieval Coroner* (Cambridge University Press, 1961) 32.

4 For further analysis on the history of the deodand, its abolition and the railway industry in Britain, see Jacob Finkelstein, 'The Goring Ox: Some Historical Perspectives on Deodands, Forfeitures, Wrongful Death and the Western Notion of Sovereignty' (1972–1973) 46(2) *Temple Law Quarterly* 169; Elizabeth Cawthorn, 'New Life for the Deodand: Coroners' Inquests and Occupational Deaths in England, 1830–46' (1989) 33(2) *American Journal of Legal History* 137; R.W. Kostal, *Law and English Railway Capitalism, 1825–1875* (Clarendon Press, 1994); Teresa Sutton, 'The Deodand and Responsibility for Death' (1997) 18(3) *The Journal of Legal*

The abolition of the deodand was a pivotal moment in the modernisation of the office of coroner in the nineteenth century. The office first appeared in England in 1194.[5] The primary duty of coroners through much of the Middle Ages was to conduct inquests on dead bodies; however, they also held inquests on treasure troves, wrecks of the sea and royal fish; received abjurations of the realm; heard confessions of felons and appeals from approvers, and managed exactions and outlawries.[6] The coroner was a powerful and influential officer of the realm, second to none other than the sheriff, whose actions became subject to the former's oversight. Their importance was concomitant to their duties, which involved collecting fines, amercements, forfeitures and deodands, and keeping other financial interests of the Crown. The expense of waging crusades in the twelfth century placed immense pressure on the sovereign to devise new avenues for raising revenue, particularly with regards to the administration of criminal law, and it was the responsibility of the coroner to assist the sheriff in directing such profits to the Crown. The medieval coroner was essentially a tax collector and their responsibilities towards the Crown were fiscal in nature. Even an inquest on a dead body, conducted following a sudden, unnatural or violent death, provided ample opportunity to collect amercements from individuals or townships.[7]

It is no accident that this book commences with the spectral image of the deodand, a legal technology that recalls the fiscal origins of the coroner. The deodand was both a material object, a thing in itself that caused the death of a

History 44; Elizabeth Cawthorn, *Job Accidents and the Law in England's Early Railway Age* (Edwin Mellen Press, 1997); Teresa Sutton, 'The Nature of the Early Law of Deodand' (1999) 30(9) *Cambrian Law Review* 14; Adrian Gray, 'A Review of Transport and the Law of Deodand' (2011) 212 *Journal of the Railway and Canal Historical Society* 26.

5 The allusion to *custos placitorum corones* ('keepers of the pleas of the Crown') can be found in Chapter XX of the *Articles of Eyre* (1194). There is some evidence, however, that in a rudimental form the office appeared during the reign of King Alfred in the ninth and tenth centuries. There is scholarly debate about whether the Articles of Eyre instituted an entirely new office or merely codified pre-existing duties that had been assumed by other officers of the realm. See R.H. Wellington, *The King's Coroner* (William Clowes & Sons Ltd, 1905); F.J. Waldo, 'The Ancient Office of Coroner' (1910–1911) 8 *Transactions of the Medico-Legal Society* 101; Sir John Jervis, *Office and Duties of Coroners* (Sweet & Maxwell Ltd, 7th edn, 1927); T.R. Forbes, 'Crowner's Quest' (1978) 68(1) *Transactions of the American Philosophical Society* 1; Jill McKeough, 'Origins of the Coronial Jurisdiction' (1983) 6 *University of New South Wales Law Journal* 191.

6 Ian Freckelton and David Ranson, *Death Investigation and the Coroner's Inquest* (Oxford University Press, 2006) 6–13. The duties of the coroner were first set out in *Officio Coronatoris: The Office of The Coroner* 4 Edward 1 AD 1275, 1276, which Hunnisett notes was actually an extract from Henri de Bracton's *De Legibus et Consuetudinibus Angliae* (Sir Travers Twiss ed, 1879), but over time assumed the form of a statute. It regulated the conduct of the English coroner until the enactment of the *Coroners Act 1887* (UK).

7 Sara M. Butler, *Forensic Medicine and Death Investigation in Medieval England* (Routledge, 2015) 3–4. For further information about the fiscal duties of the medieval coroner, see Charles Gross (ed), *Select Cases from the Coroner's Rolls, A.D. 1265–1413* (Bernard Quaritch, 1896).

person, and a juridical device, deployed by coroners to cultivate a monetary relationship between the living and the dead. While its abolition in the mid-nineteenth century consigned many of the coroner's financial duties to the tomes of legal history and replaced a fiscal coroner with a medico-legal coroner, the latter of whom became solely preoccupied with investigating the causes of sudden, unnatural or violent deaths, the deodand exemplified the important role that technology assumed in the death investigation process. This book examines how different technologies affected the way coroners occupied their office in the nineteenth and twentieth centuries. The nineteenth-century coroner was quite unlike the medieval knight who was elected by county freeholders to 'keep watch over the profits of the Crown' in the thirteenth century.[8] By the turn of the twentieth century, the coroner had become preoccupied with form-filling, record-keeping and other administrative tasks.

The modernisation of the office of coroner took place amid significant political, social and economic transitions in Britain. The Industrial Revolution was a period of rapid changes in mechanised transportation, machine-based production and manufacturing technology that occurred from the 1760s to 1840s. This epoch saw the mass movement of proletariat populations from small agricultural towns to large manufacturing cities, the liberalisation of trade and resultant emergence of a market-based economy, and the rise of bureaucratic governance and the administrative nation-state.[9] What distinguished industrial capitalism from earlier forms of acquisition, according to Max Weber, was 'the pursuit of profit, and forever *renewed* profit, by means of continuous, rational, capitalist enterprise'.[10] And what conditioned this unrelenting accumulation, but also reinvestment of capital, was 'the rational capitalistic organization of (formally) free labour'.[11] However, Weber further explains that the separation of factories from households, the invention of 'double-entry book-keeping', the technical calculability of knowledge, the rational codification of law and Protestant ethic were also all historically specific to the emergence of modern capitalism in the West.[12] The work of the coroner was not immune from the widespread

---

8 Wellington, above n 5, 6.
9 See, for example, Frederick Engels, *The Condition of the Working-Class in England in 1844* (Florence Kelley Wischnewetzky trans, George Allen & Unwin Ltd, 1892); David Harvey, *The Urbanization of Capital: Studies in the History and Theory of Capitalist Urbanization* (Johns Hopkins University Press, 1985); Immanuel Wallerstein, *The Modern-World System IV: Centrist Liberalism Triumphant 1789–1914* (University of California Press, 2011).
10 Max Weber, *The Protestant Ethic and the Spirit of Capitalism* (Talcott Parsons trans, Routledge, 1992) xxxii.
11 Ibid, xxxiv (emphasis in original).
12 Ibid, xxxix. It is important, as Georges Bataille writes, not to reduce capitalist enterprise to the 'restrictive economy' of 'historical materialism'. Discourses of waste, excess and consumption are as important as paradigms of production, growth and accumulation for understanding a history of capitalism in the West. Bataille identifies excess as the accursed share of a general economy, which must be exuded, spent, squandered, to avoid, for example, its inundation in times of war. Interestingly, Bataille describes eating, sex and *death* as the ultimate

changes shaped by the industrialisation of Britain in the eighteenth and nineteenth centuries. The provision of an annual salary to coroners, the payment of medical witnesses and jurors for their attendance at inquests, the formation of coronial societies and professional associations, and the emergence of medical jurisprudence as a specialised field of expertise were all hallmarks of the late nineteenth century.

In *The History of Sexuality, Volume 1*, Michel Foucault traces a shift in the eighteenth century from the sovereign's 'right to decide life and death' towards a 'power of life and death'.[13] The latter was more concerned with techniques for managing life, maximising its efficacy and exploiting its vitality, evinced by a governmental preoccupation with monitoring fluctuations in birth and death rates. Foucault calls this new relation of power–knowledge 'biopower', which consisted of an *anatomo-politics* of the body and a *bio-politics* of the population.[14] The regulation of death, including its control by the state and a range of medical, legal and financial institutions, also gave rise to a new arrangement of *thanato-politics*, or the governance of the dead in the West.[15] For Foucault, 'government' does not simply refer to an institution, but to a historical practice that had as its ultimate aim the care of a population. In the eighteenth and nineteenth centuries, the art of governing developed a range of techniques for managing populations of the dead as well as the living.[16]

Changes to the means of production, specifically the transition from agricultural to machine-based labour, produced devastating, often fatal consequences for workers and their families during the Industrial Revolution. This resulted in both disdain for coroners, particularly among capitalists and their supporters due to an increase in the number of inquests conducted on workplace deaths, and reverence for coroners from the relatives of the deceased and communities in general for holding employers accountable for causing sudden, unnatural or violent deaths. Indeed, coroners were often venerated for admonishing capitalists for creating dangerous workplaces, recommending changes to industrial practices and affirming legal responsibility for occupational health and safety, which were all oriented towards the governmental objective of improving the health of the population. The office of coroner should be viewed, then, as what Foucault

---

luxuries of accumulation, an excess that reveals the fragility of a restrictive economy of capitalism: Georges Bataille, *The Accursed Share: An Essay on General Economy – Volume 1: Consumption* (Robert Hurley trans, Zone Books, 1991).

13 Michel Foucault, *The History of Sexuality, Volume 1: The Will to Knowledge* (Robert Hurley trans, Penguin Books, 1998) 136.

14 Ibid, 139.

15 Michel Foucault, 'The Political Technology of Individuals', in Luther H. Martin, Huck Gutman and Patrick H. Hutton (eds), *Technologies of the Self: A Seminar with Michel Foucault* (Tavistock, 1988) 160.

16 See generally, Zohren Bayatrizi, *Life Sentences: The Modern Ordering of Mortality* (University of Toronto Press, 2008); Finn Stepputat (ed), *Governing the Dead: Sovereignty and the Politics of Dead Bodies* (Manchester University Press, 2014).

calls an apparatus (*dispositif*),[17] which, in the nineteenth century, had as its remit the ordering of mortality, alongside other legal, medical and bureaucratic institutions. The coroner was positioned as a conduit of a new power–knowledge network, at the nexus of the nation-state's burgeoning interest in managing relations between the living and the dead.

The role that coroners performed during the development of industrial capitalism was not confined to Britain. The office was disseminated throughout the British colonies in the eighteenth and nineteenth centuries through a forceful expansion of the imperial project of empire building.[18] While the funding model, appointment procedures and administrative operations differed on a local level between the various colonies, the way the office cultivated relations between the living and the dead was similarly tied to legitimating imperial rule. In holding inquests on the deaths of colonists – though also less frequently on the remains of Indigenous persons[19] – coroners not only disavowed the investigatory rituals of Indigenous cultures, but also authorised the appropriation of Indigenous lands as lawful. What I am gesturing towards here is the idea that the inquest was a palimpsestic technology, which is to say that it inscribed the lands with the remains of colonial settlers and concealed traces of the Indigenous ancestors that indwelled within. I am not suggesting that the coroner was an infallible, unnerving cog in the machinery of empire building, but rather that their role was part of the modalities of colonial governance.

This book offers a historical and critical examination of the relationship between a legal institution and the governance of the dead in Australia. The focus on the Australian colonies provides a unique opportunity to historicise the

---

17   What I'm trying to single out with this term is, first and foremost, a thoroughly heterogeneous ensemble consisting of discourses, institutions, architectural forms, regulatory decisions, laws, administrative measures, scientific statements, philosophical, moral and philanthropic propositions – in short, the said as much as the unsaid. Such are the elements of the apparatus. The apparatus itself is the system of relations that can be established between these elements.
Michel Foucault, 'The Confession of the Flesh', in Colin Gordon (ed), *Power/Knowledge: Selected Interviews and Other Writings 1972–1977* (Pantheon Books, 1980) 194

18   For legal histories of colonialism and the British Empire, see Diane Kirkby and Catharine Coleborne (eds), *Law History Colonialism: The Reach of Empire* (Manchester University Press, 2001); Shaunnagh Dorsett and Ian Hunter (eds), *Law and Politics in British Colonial Thought: Transpositions of Empire* (Palgrave Macmillan, 2010); Ian Duncanson, *Historiography, Empire and the Rule of Law: Imagined Constitutions, Remembered Legalities* (Routledge, 2012); Shaunnagh Dorsett and John McLaren (eds), *Legal Histories of the British Empire: Laws, Engagements, Legacies* (Routledge, 2014); Prem Kumar Rajaram, *Ruling the Margins: Colonial Power and Administrative Rule in the Past and Present* (Routledge, 2014).

19   Shaun McVeigh explains that, '[p]rior to the 1830s, the common law applied to settlers because of their status as subjects. It did not apply to Indigenous Australians precisely because they were not yet considered to be subjects': 'Law As (More or Less) Itself: On Some Not Very Reflective Elements of Law' (2014) 4 *University of California Irvine Law Review* 471, 478.

technologies of the death investigation during a turbulent, brutal epoch of empire building. The appropriation of Indigenous land and the oppression of Indigenous Australians, the tyranny of distance and the exploitation of natural resources all render this period of transition from colonial administration to democratic representation in Australia in the nineteenth and twentieth centuries a unique context to examine the formation of legal relations between the living and the dead. While writing a history of the Australian coroner is a significant project in itself, by offering tools for thinking institutionally about law, I hope this history can become paradigmatic of a broader institutional relationship between law and the dead. I emphasise in each chapter that the key to making sense of the manifold ways in which coroners encountered the dead in the nineteenth and twentieth centuries is by conceptualising law as a network of institutions, relations and technologies. In light of this theoretical framework, each chapter explores a different technology – place-making, the forensic gaze, bureaucratic manuals, record-keeping and radiography – so as to give an account of how the living formed legal relations with the dead, but also how they continue to manage those relations today. The pages that follow should not only be read as a history of the past, but as Foucault famously writes, a 'history of the present'.[20]

Thinking institutionally

In *Discipline and Punish*, Foucault historicises the birth of the prison by theorising the modalities of institutions, which have at their disposal specific technologies for investing the body 'to carry out tasks, to perform ceremonies, to emit signs'.[21] In *The History of Sexuality, Volume 1*, he critiques discourses of sexuality by examining a range of institutions in their concrete arrangements – that is, in 'a continuum of apparatuses'.[22] In other words, key to understanding Foucault's writings on prisons, sexuality, medicine and the human sciences is his conceptualisation of institutions as material and discursive functionings in a network of relations. The importance of institutional thinking in his thought may have been inspired by Giambattista Vico's assertion that '[t]he order of ideas must proceed from the order of institutions'.[23] Here, institutions appear as that which attach human beings not only to laws, but also to places, traditions and relations.

In an essay titled 'Nietzsche, Genealogy, History', Foucault sets out a new methodology for critical historicism that is galvanised by his investigations into the plurality of institutional life. Genealogy, as a form of writing history, does

---

20 Michel Foucault, *Discipline and Punish: The Birth of the Prison* (Alan Sheridan trans, Penguin Books, 1991) 31.
21 Ibid, 25.
22 Foucault, above n 13, 144.
23 Giambattista Vico, *New Science* (David Marsh trans, Penguin Books, 3rd edn, 1999) §238.

not excavate the origins of things, but finds temporal disparity in the descent of things:

> to follow the complex course of descent is to maintain passing events in their proper dispersion; it is to identity the accidents, the minute deviations – or conversely, the complete reversals – the errors, the false appraisals, and the faulty calculations that gave birth to those things that continue to exist and have value for us.[24]

Distilled into its essential components, Paul Veyne surmises, genealogical analysis invokes a history of relations.[25] Relations are irreducible to institutions, yet they are constitutive of institutional practices. In this vein, this book suggests that law cannot be thought of outside its institutional relations, formations and arrangements. It engages in what Judith Butler calls the *modus operandi* of critique, but one which could also equally apply to the practice of genealogy: '[c]ritique is always a critique of some instituted practice, discourse, episteme, institution, and it loses its character the moment in which it is abstracted from its operation'.[26]

Law and the Dead offers readers a genealogy of the transformations of a particular type of legal institution in the nineteenth and twentieth centuries. It thus provides tools for historicising the way in which the coronial institution, in the pursuit of investigating a sudden, unnatural or violent death, assumed responsibility for caring for the dead. In 'The Political Technology of Individuals', Foucault recounts how 'caring' for individual life emerged as 'a duty for the state' in the eighteenth century.[27] I suggest that caring for the dead also appeared as an object of governance during the modern era. However, where I diverge from Foucault is in relation to his emphasis on governance as a *political* technology. Indeed, this book conceives of practices for taking care of the dead as more specific, and perhaps more local, than techniques of governmentality. It conceptualises care as a way to maintain legal relations with the dead. Thinking like this asks readers to conceive of law in terms of the materiality of its institutions, the technologies that attach themselves to institutional practices and the performances of office that sustain the vitality of legal institutions.

In writing an institutional history of law and the dead, a history that emphasises law's institutional formations, this book pays attention to questions of technology, for technologies comprise the repertoire of institutional practice. The

---

24 Michel Foucault, 'Nietzsche, Genealogy, History', in David F. Bouchard (ed), *Language, Counter-Memory, Practice: Selected Essays and Interviews* (Cornell University Press, 1977) 146.

25 Paul Veyne, 'Foucault Revolutionizes History', in Arnold I. Davidson (ed), *Foucault and His Interlocutors* (Catherine Porter trans, The University of Chicago Press, 1997) 177.

26 Judith Butler, 'What Is Critique? An Essay on Foucault's Virtue', in David Ingram (ed), *The Political Blackwell Readings in Continental Philosophy* (Wiley-Blackwell, 2002) 1.

27 Foucault, above n 15, 147.

term 'technology' derives from the Ancient Greek word for technique (*technê*), which denotes a skill, art or craft. *Technê* was conceived of as an implement, tool or instrument capable of producing 'a mechanical, physical, or chemical effect'.[28] It involved a skilful, operational or industrial process, action or practice, which could only be harnessed by human 'technician[s]'.[29] This book historicises coroners as occupying a civic role, who, with the assistance of a range of technologies, were capable in the nineteenth and twentieth centuries of forming and shaping relations properly belonging to law. While the technologies examined in subsequent chapters are paradigmatic of how coroners occupied their office, I am not suggesting that they are the only technologies capable of holding on to law's institutional life, transforming legal attitudes or modifying the conduct of legal relations with the dead. Rather, each technology considered herein exemplifies how the dead became attached to the institutional life of law and how coroners assumed a responsibility to take care of the dead. This is useful for drawing attention to a particular form of writing a history of the death investigation process, one that pays attention to how law operates institutionally, and constitutes a genealogy of the transformations in the way the state and its officials formed legal relations with the dead in the nineteenth and twentieth centuries.

Legal scholars have recently questioned what it means to think about law archivally – that is, to question the 'ethical, aesthetic, and emotional aspects of archives and the demands that law's materials make upon scholars to exercise care and judgment'.[30] This 'archival turn' in legal scholarship follows an earlier turn in the humanities and social sciences. Historians, anthropologists and literary scholars have long questioned 'conventional modes of writing history, highlighting the (im)possibility of recuperating historical and archival texts as "truth" and urging the need to employ critical and literary modes of reading'.[31]

---

28 Marcel Mauss, 'Technology (1935/1947)' (Dominique Lussier trans) in *Techniques, Technology and Civilisation* (Durkheim Press, 2006) 98. On the link between technology and technique, see Andrew Barry, *Political Machines: Governing a Technological Society* (Athlone Press, 2001).

29 Ibid, 114.

30 Katherine Biber and Trish Luker, 'Evidence and the Archive: Ethics, Aesthetics, and Emotion' (2014) 40(1) *Australian Feminist Law Journal* 1, 14.

31 Renisa Mawani, 'Law's Archive' (2012) 8 *Annual Review of Law and Social Science* 337, 340. Since the early 2000s, there has been a proliferation of publications on archival history: Carolyn Hamilton et al. (eds), *Refiguring the Archive* (Springer, 2002); Carolyn Steedman, *Dust: The Archive and Cultural History* (Rutgers University Press, 2002); Antoinette Burton (ed), *Archive Stories: Facts, Fiction and the Writing of History* (Duke University Press, 2005); Annelise Riles (ed), *Documents: Artifacts of Modern Knowledge* (University of Michigan Press, 2006); John Ridener, *From Polders to Postmodernism: A Concise History of Archival Theory* (Litwin Books, 2008); Laura Ann Stoler, *Along the Archival Grain: Epistemic Anxieties and Colonial Common Sense* (Princeton University Press, 2010); Arlene Farge, *The Allure of the Archives* (Thomas Scott-Railton trans, Yale University Press, 2015); Niamh Moore, Andrea Salter, Liz Stanley and Maria Tamboukou, *The Archive Project: Archival Research in the Social Sciences* (Routledge, 2016).

In the legal discipline, 'archival thinking' problematises the way legal historians have approached archives as a source of truth or a site of enlightenment about the origins of law. But, further than this, it has asked legal historians to reflect on the 'allure and anxieties' of 'law's archives', and to question the ethics of our own engagement with its records. The question that arises in working archivally with law is therefore one of scholarly conduct. It raises the problem of establishing an ethical relation to the uses of archival research.

Thinking archivally also involves conceptualising law's archives as polycentric. It reveals that every legal archive is marked by infinitesimal acts of curation. The archive can be described as Keith Jenkins says of history in general: it is 'a literary narrative *about* the past, a literary composition of the data into *a* narrative where the historian creates *a* meaning *for* the past'.[32] The chapters that follow *narrate* a coronial archive from newspapers, blueprints, rolls, registers, memos, circulars, documents and manuals. Yet this specific archive, like all archives, remains incomplete, inexhaustible and uncertain, particularly for what it fails to record: *the unknown dead*. Here, I refer not only to deaths that were not reported to or investigated by coroners, inquests that were not recorded and correspondence that was lost along the way, but also to the names of the dead that were forgotten, were never identified and will remain forever absent. These names belong predominantly to women, non-Europeans and the poor, and they are redolent of many untold stories from the nineteenth and twentieth centuries of botched abortions, suicides, hate crimes, industrial accidents, deaths in custody and so on. To reflect, then, on how to form an ethical relationship to the coronial archive, legal scholars must take responsibility for its incompleteness and the subjectivity of academic curation. Institutional histories must do more than describe what happened in the past; they should interrogate the excesses of the archive, the 'marginalia, transgressions between official and unofficial records, rumour, the unexpected, unwritten, and the unsaid'.[33]

Finally, this book makes an intervention into the historiography of death, which emerged as a prominent field of social history in the 1970s.[34] Death as an object of study has come to prominence in the fields of sociology, anthropology, psychology

---

32 Keith Jenkins, *Re-thinking History* (Routledge Classics, 2003) xii (emphasis in original).
33 Biber and Luker, above n 30, 9.
34 Since then, there have been a proliferation of social and cultural histories of death: see, e.g., Joachim Whaley (ed), *Mirrors of Mortality: Studies in the Social History of Death* (Routledge, 1981); Clare Gittings, *Death, Burial and the Individual in Early Modern England* (Routledge, 1984); Olive Anderson, *Suicide in Victorian and Edwardian England* (Clarendon Press, 1987); Patrick J. Geary, *Living with the Dead in the Middle Ages* (Cornell University Press, 1994); Pat Jalland, *Death in the Victorian Family* (Oxford University Press, 1996); Bruce Gordon and Peter Marshall (eds), *The Place of the Dead: Death and Remembrance in Late Medieval and Early Modern Europe* (Cambridge University Press, 2000); Pat Jalland, *Australian Ways of Death: A Social and Cultural History, 1840–1918* (Oxford University Press, 2002); Allan Kellehear, *A Social History of Dying* (Cambridge University Press, 2007).

and theology since at least the late nineteenth century.[35] Indeed, the social sciences have informed histories of death, especially through the psychoanalytic notion that its denial is a natural quality of being human, and they have contributed to what I term a 'repressive hypothesis' of death, most eloquently articulated by the French historian, Philippe Ariès.[36] According to Ariès, transformations in Western attitudes towards death can be categorised according to four temporally distinct stages: tamed death, the individual death, the death of the other and the denial of death. Whereas the taming of death in the Middle Ages fostered a close intimacy between the living and the dead, the eighteenth century projected death as an 'unaccepted separation' between the self and other; and by the late nineteenth century, death had become forbidden, '*unnameable*' and sequestered in medical and legal institutions.[37] In the final chapter of *The Hour of Our Death*, Ariès argues that institutionalisation of the dead in the nineteenth century progressed to its banishment in the twentieth century, when death was refused and the dead were hidden (or silenced) in hospitals, morgues and cemeteries.[38] For Walter Benjamin, this historical process took place much earlier in the West:

> It has been observable for a number of centuries how in the general consciousness the thought of death has declined in omnipresence and vividness. … Dying was once a public process in the life of the individual and a most exemplary one … In the course of modern times dying has been pushed further and further out of the perceptual world of the living.[39]

---

35 See, e.g., Émile Durkheim, *Suicide: A Study in Sociology* (John A. Spaulding and George Simpson trans, Routledge, 2005); Joseph Jacobs, 'The Dying of Death' (1899) 72 *Fortnightly Review* 264; Robert Hertz, *Death and the Right Hand* (Rodney Needham and Claudia Needham trans, Cohen & West, 1960); Herman Feifel, *The Meaning of Death* (McGraw-Hill, 1st edn, 1959); Jessica Mitford, *The American Way of Death* (Simon and Schuster, 1st edn, 1963); Geoffrey Gorer, *Death, Grief and Mourning in Contemporary Britain* (Cresset Press, 1965).

36 The 'denial of death' was first articulated by Sigmund Freud in 'Thoughts for the Times on War and Death' and 'Mourning and Melancholia', in Sigmund Freud, *On the History of the Psycho-Analytic Movement, Papers on Metapsychology and Other Works, Volume XIV (1914–1916)* (James Strachey trans, Vintage, The Hogarth Press and the Institute of Psycho-Analysis, [1915] 2001 edn). While his conceptualisation of a death instinct in 'Beyond the Pleasure Principle' is more complex, this has not prevented sociologists and psychologists from formulating a general theory of the repression of death in the unconscious: see Sigmund Freud, *Beyond the Pleasure Principle, Group Psychology and Other Works, Volume XVII (1920–1922)* (James Strachey trans, Vintage, The Hogarth Press and the Institute of Psycho-Analysis, 2001); Elizabeth Kübler-Ross, *On Death and Dying* (Simon and Schuster, 1969); Ernest Becker, *The Denial of Death* (Free Press Paperbacks, 1973).

37 Philippe Ariès, *Western Attitudes toward Death: From the Middle Ages to the Present* (Patricia M. Ranum trans, The Johns Hopkins University Press, 1974) 106 (emphasis in original).

38 Philippe Ariès, *The Hour of Our Death: The Classic History of Western Attitudes Toward Death Over the Last One Thousand Years* (Helen Weaver trans, Vintage Books, 2nd edn, 2008).

39 Walter Benjamin, *Illuminations: Essays and Reflections* (Harry Zohn trans, Schocken Books, 1968) 93–94.

The repressive hypothesis dominated historiographies of death until the 'critical turn' within death studies in the 1990s. Refuting the thesis that death disappeared, as well as the writings of Sigmund Freud, Geoffrey Gorer and others, contemporary historians, sociologists and anthropologists have claimed that an incitement to discourse on death never declined in the modern era.[40] Challenging the notion that the dead were sequestered from the living by the late nineteenth century, scholars have argued that death has been both absent and present across all time periods: 'it surfaces and submerges but never disappears'.[41] Despite this critical turn, as Jonathan Dollimore points out, the repressive hypothesis persisted in historical, sociological and philosophical treatises on death towards the end of the twentieth century.[42] Even in *The Revival of Death*, one of the most prominent works associated with the birth of 'death studies', the emphasis on a revival presupposes a historical prohibition against discourses of death. Indeed, Tony Walter argues that rational bureaucracy was chiefly responsible for removing death from the public sphere and hiding it in private spaces. Conversely, *Law and the Dead* offers a provocative challenge to conventional thinking about the sequestration of the dead in the nineteenth and twentieth centuries. It argues that the persistence of the repressive hypothesis, or rather its oblique revival in scholarship that encourages society to talk more about death, arises from a misrecognition of the institutional forms that the dead occupied in the past and continue to assume in the present.

While social histories of death have provided tools for studying how the living engaged with the dead in different epochs, advocates of the hypothesis that death was denied in the late nineteenth century often render institutions from that period monolithic, despotic and lifeless. They ignore the plurality and vitality of institutional practices in the modern era, and their productive capacity in forming and cultivating relations with the dead. There has always

---

40 The most notable works associated with the 'critical turn' within death studies are Jenny Hockey, *Experience of Death: An Anthropological Account* (Edinburgh University Press, 1990) and Tony Walter, *The Revival of Death* (Routledge, 1994). See also David Clark (ed), *The Sociology of Death: Theory, Culture and Practice* (Blackwell, 1993); Philip A. Mellor and Chris Shilling, 'Modernity, Self-Identity and the Sequestration of Death' (1993) 27(3) *Sociology* 411; Glennys Howarth (ed), *The Changing Face of Death: Historical Accounts of Death and Disposal* (Macmillan Press, 1996); Elizabeth Hallam, Jenny Hockey and Glennys Howarth, *Beyond the Body: Death and Social Identity* (Routledge, 1999); Allan Kellehear, *Death and Dying in Australia* (Oxford University Press, 2000); Glennys Howarth, *Death and Dying: A Sociological Introduction* (Polity, 2007); Elizabeth Hallam and Jenny Hockey, *Death, Memory and Material Culture* (Berg Publishers, 2010).

41 Warren Smith, 'Organizing Death: Remembrance and Re-collection' (2006) 13(2) *Organization* 225, 226.

42 Jonathan Dollimore, *Death, Desire and Loss in Western Culture* (Routledge, 1998) 119–123. Examples of this include Zygmunt Bauman, *Mortality, Immortality and Other Life Strategies* (Polity Press, 1992) and Jean Baudrillard, *Symbolic Exchange and Death* (Iain Hamilton Grant trans, Sage Publications, 1993). For Baudrillard the entire 'rationality' of Western culture is organised around 'the exclusion of the dead and of death': Ibid, 126.

been life in legal institutions, as this book will make clear; and, in revealing the expression of their vitality, it depicts institutions teeming with the chatter of the dead. To put this differently, discourses of death were not repressed *per se* by medical and legal institutions in the nineteenth and twentieth centuries; rather, they were re-signified, re-conceptualised and transfigured, as part of a process of which denial was but one element of speaking with and relating to the dead.[43]

This book also makes an intervention in historiographies of death by historicising institutions of *the dead*. It is all too common to find in the literature a conceptual slippage between the syntax of the dead and death. For example, in *Western Attitudes towards Death*, Ariès often fails to distinguish between the materiality of the dead and the occurrence of death. This is further apparent in philosophical treatises on the relationship between language and death as well as metaphysical reflections on the impossible experience of death[44] – though an exception to this general trend is Thomas Laqueur's monumental tome, *The Work of the Dead*, where the dead are positioned as isomorphic to the corpse. As Laqueur claims, 'the dead body matters, everywhere and across time, as well as in particular times and particular places'.[45] While this book often deploys the language of the corpse and of the dead interchangeably, my intention is not to reduce the latter to the decaying flesh of mortal remains. The work of the dead was far more pervasive in the nineteenth century than the final resting place of the corpse. Its institutional presence was far more penetrating than the inscription of names on the walls of monuments.

It is only by holding on to the dead as a historical object in itself and historicising the institutional relations it forms that one can interrogate

---

43 Lindsay Prior has made a similar point in *The Social Organisation of Death: Medical Discourse and Social Practices in Belfast* (Macmillan, 1989).

44 There are too many references to mention here. However, I consider the following to be the main texts in a 'hermeneutics of death': Jacques Choron, *Death and Western Thought* (Collier Books, 1963); Giorgio Agamben, *Language and Death: The Place of Negativity* (Karen E. Pinkus and Michael Hardt trans, University of Minnesota Press, 1991); Georges Bataille, *Theory of Religion* (Robert Hurley trans, Zone Books, 1992); Jacques Derrida, *Aporias* (Thomas Dutoit trans, Stanford University Press, 1993); Jacques Derrida, *The Gift of Death* (David Wills trans, The University of Chicago Press, 1995); Michel Foucault, *Death and the Labyrinth* (Charles Ruas trans, Continuum, 2004); Michel Foucault, *Speech Begins after Death: In Conversations with Claude Bonnefoy* (Robert Bononno trans, University of Minnesota Press, 2013). On hauntology as a methodology of literary studies, see Jacques Derrida, *Specters of Marx: The State of the Debt, the Work of Mourning and the New International* (Peggy Kamuf trans, Routledge, 2006); Jacques Derrida et al., *Ghostly Demarcations: A Symposium on Jacques Derrida's Specters of Marx* (Verso, 2008); María del Pilar Blanco and Esther Peeren (eds), *The Spectralities Reader: Ghosts and Hauntings in Contemporary Cultural Theory* (Bloomsbury, 2013).

45 Thomas W. Laqueur, *The Work of the Dead: A Cultural History of Mortal Remains* (Princeton University Press, 2015) 1.

questions of law.[46] Indeed, to write an institutional history of coronial law, I suggest that one must disavow the temptations to liberate death from the conceptual shackles of its banishment in the past, and interrogate the dispositions of the dead, and the conduct, forms and roles the dead have assumed throughout history. If this hypothesis is accepted, it follows that the collapsing of the language of death with that of the dead may account for the conspicuous absence of legal institutions in historiographies of death. This book intervenes in this field of inquiry by questioning where legal institutions fit in a social and cultural history of death, and what is made possible by taking seriously the important role assumed by coroners in the governance of the dead.[47] It thus makes a unique contribution by asking historians to take seriously the question of legal relations, while asking legal scholars to trace the historical formations between institutions and the dead.

Trajectory of the book

Chapter 1, 'Law in the necropolis', begins by tracing a spatial history of the death investigation process in the nineteenth century. The chapter offers a historical account of how the movements of the coroner incorporated the dead into the institutional life of coronial law. The proximity of the dead to the living emerged as a spatial problem in Australia in the late eighteenth and early nineteenth centuries. Not only did the unburied corpse pose a danger to the physical health of urban dwellers, but its presence was believed to breach the boundaries between the realms of the living and the dead. Coroners were appointed by colonial governments to collect, identify and investigate the remains of the dead, but also, importantly, to remove the unburied corpse from the public

46 The category of the dead has long been subject to doctrinal analysis in the legal discipline. Indeed, legal scholars have questioned whether a corpse is a person or a thing in law. If it is held to be a thing, they have asked what obligations, if any, pertain to owning the dead; while, if it is held to be a person, they have queried whether the dead possess rights and, if so, whether those rights should survive a death. The literature on 'rights' spans many different areas of law, such as succession law and the law of estates, health and medical law, and human rights law. For a more recent examination of this topic, see Desmond Manderson (ed), *Courting Death: The Law of Mortality* (Pluto Press, 1999); Ray Madoff, *Immortality and the Law: The Rising Power of the American Dead* (Yale University Press, 2010); Norman L. Cantor, *After We Die: The Life and Times of the Human Cadaver* (Georgetown University Press, 2010); Heather Conway, *The Law and the Dead* (Routledge, 2016). On questions of repatriation of Indigenous remains, see Cressida Fforde, Jane Hubert and Paul Turnbull (eds), *The Dead and Their Possessions: Repatriation in Principle, Policy and Practice* (Routledge, 2002); Paul Turnbull and Michael Pickering (eds), *The Long Way Home: The Meaning and Values of Repatriation* (Berghahn Books, 2010).

47 Historiographies of nineteenth- and twentieth-century coroners have largely been written by social, medical and cultural historians. See, e.g., Michael Clark and Catherine Crawford (eds), *Legal Medicine in History* (Cambridge University Press, 1994); Ian Burney, *Bodies of Evidence: Medicine and the Politics of the English Inquest, 1830–1926* (Johns Hopkins University Press, 2000).

sphere. However, the manner in which they moved, the way they carried the dead through the streets of the city, how they set up inquests in taverns and stored remains in outhouses revealed the different ways in which they formed legal relations with the dead. The itinerant coroner harnessed a range of jurisdictional technologies, such as walking, hawking and building, to establish a lawful place for the dead in the city.

The post-mortem examination was a compulsory procedure of the death investigation process from the 1860s and 1870s. Prior to this, autopsies were seldom conducted before a coronial inquest; it would suffice for the coroner to merely glance over the surface of the corpse. The development of medical jurisprudence as a sub-discipline of human anatomy transformed the way coroners and jurors viewed the corpse. It positioned the forensic gaze as intrinsic to the legal custom of *super visum corporis*, which meant that an inquest could only take place 'upon view of the dead body'. Chapter 2, 'Visual regimes of the dead', offers a historical account of the transformations in the techniques used for viewing the dead body in the late nineteenth century. The chapter argues that changes to the forensic gaze were integral to how coroners maintained legal relations with the dead. The gaze was part of a repertoire of jurisdictional technologies through which the coroner assumed responsibility for taking care of the dead.

In Chapter 3, 'The bureaucratic logic of office', I examine how the coronial manual functioned as a technology of office in the late nineteenth and early twentieth centuries. The manual guided coroners in interpreting the scope of their jurisdiction, the performance of their duties and the proper administration of inquests. It technocratised the practices of coronial law and procedure, while offering guidance on how to fulfil the obligations of the coroner's office. The manual was undoubtedly preoccupied with questions of technical knowledge, skill and expertise; yet this does not mean that it was bereft of an ethics of responsibility. It held on to the question of responsibility by framing the death investigation process within a bureaucratic logic of office. Chapter 3 thus shows how the coronial manual assumed an indispensable role in the formation of an ethical mindset towards the dead.

While Chapter 3 traces a bureaucratic logic in the transformations of the coronial manual and its inventory of circulars, forms and precedents in the late nineteenth and early twentieth centuries, Chapter 4, 'Dead records', investigates how records, files and documents, created, collected, submitted and reproduced in the performance of office, shaped the bureaucratic impulse of the coronial jurisdiction. The chapter explores how record-keeping became an integral part of the modernisation of the coroner's court at the turn of the twentieth century. It first argues that the conduct of the coroner's office was institutionalised as a court of record, a sitting of the coroner's court, through the technology of the file. But it also considers the effects of this technology on both the role of the coroner, who increasingly assumed responsibility for recording a biography of the dead, and on the dead, who appeared as neither things nor persons, but records in an archive of institutional memory. Chapter 4 concludes by suggesting that the duty of record-keeping was essential to how coroners took care of the dead. Coroners collected biographical information about the deceased, narrating their lives and deaths, through the technology of the file, which has come to signify one of the most important functions of the coronial jurisdiction.

The final chapter of the book, 'Screening the corpse', examines how forensic radiography transformed the way coroners took care of the dead in the twentieth century. The use of X-rays during the death investigation process altered the coroner's view of the corpse, a view that was necessary in the late nineteenth century for an inquest to proceed. By the mid-twentieth century, the coronial jury ceased to view the corpse during an inquest, while the post-mortem examination was left almost entirely in the hands of the forensic pathologist. The invention of radiography 'mechanised' the forensic gaze that evolved in the nineteenth century; that is to say, medico-legal representations of the corpse no longer relied upon the cutting up of bodies. The coroner's view of the corpse became increasingly mediated by medical imaging technology. Far from interpreting the mechanised gaze as one that detached the corpse from the institutional life of coronial law, this chapter argues that it transformed the quality of caring practices. Coroners had to develop a new aptitude for viewing the interiority of the corpse from afar, and had to acquire new skills for interpreting the significance of shadows as evidence of death causation. The history of forensic radiography demonstrates how new technologies continue to make demands on coronial institutions to think differently about how they cultivate legal relations with the dead.

The technologies analysed in each chapter are historically significant in framing our understanding of how the office of coroner was conducted, how the duties of that office were performed, and how the technologies that corresponded to those duties shaped legal relations in the nineteenth and twentieth centuries. Each chapter provides tools for tracing a genealogy of the institutional life of coronial law, but also for mapping how the dead have come to shape the traditions, customs and procedures of law. This requires that we take seriously how a particular type of legal institution, which underwent significant change in the nineteenth and twentieth centuries, developed technologies for managing legal relations between the living and the dead, and how in the performance of a civic role, judicial officers assumed responsibility to take care of the dead. This book thus contends that legal scholars need to think through and with institutions, relations and technologies when theorising the ethical, political and juridical problems that arise from the presence of the dead in the world.

1 Law in the necropolis

It should not have been surprising for readers of *The Argus* newspaper to hear on Thursday, 11 November 1852, that another corpse had suffered an arduous journey towards the site of an inquest in the city of Melbourne. The dead body was discovered on Monday afternoon in the town of Prahran. Witnesses reported seeing the deceased suddenly stumble and collapse, presumably due to intoxication, inside a property belonging to another man. The district coroner ordered a police constable to transport the corpse into the city for the purpose of holding an inquest, but the constable found it difficult to secure a place to store the body until the hearing could be held. He carried the corpse from one public house to another, and, after a succession of rebuffs from disgruntled innkeepers, it was finally accepted by the landlord of the Queen's Arms on Swanston Street. The partially decomposed body was stored in the cramped tavern for two days until an inquest could be held before the city coroner. The coronial jury, assembled by the coroner from an array of inebriated patrons at the Queen's Arms, swiftly returned a verdict of 'Found Dead'.

What differentiated this ordeal from other inquests reported on that day – one was conducted on the body of Fisher, who died on her way to Melbourne from the Bendigo goldfields, and another was conducted on the remains of Cunningham, who died on a schooner heading to Sydney – and what would have piqued the curiosity of *The Argus* readers, was a demand made by a Justice of the Peace, following the inquest, for the appalling incident to be investigated:

> The Bench were of opinion there existed a great want of a place in this city where dead bodies could be placed until interred, a species of *Morgue*, as used in France, as through the crowded state of the city there was not actually room to be obtained in the different public houses where a corpse could remain. Some great dereliction of duty has occurred somewhere which demands an inquiry, how a dead body should be allowed to remain from Monday until Wednesday before an inquest could be held, and in a house crowded with lodgers. We very much question whether the publicans refusing admittance to the corpse cannot be called to account, but we

do not wonder at their evincing unwillingness to receive a dead body, when they are likely to be annoyed with it, as in this case.[1]

It was unclear whether the Justice was suggesting that a 'dereliction of duty' lay with publicans, who routinely refused to accommodate the dead in their lodgings; with the city coroner, who often took a couple of days to travel to the city from his seaside manor, or with the colonial government, who failed to provide adequate facilities for conducting inquests. Indeed, *The Argus* reporter seemed to criticise publicans for refusing to accept corpses into their taverns, yet also sympathised with them given the financial consequences, not to mention social implications, that a decomposing corpse would have for sites of imbibition. But for those newspaper readers who had long been sharing their dwellings with the dead, the answer to the question of who was derelict in their duties was deceptively simple, and the construction of a deadhouse to remove corpses from the public sphere was imperative.

The movements of coroners in the nineteenth century paint a grim portrait of the plight of the dead. When a corpse was found on a public street, coroners would carry it from one public house to another, hoping to find a hospitable innkeeper willing to let a room for holding an inquest. When a dead body was discovered in a prison or hospital, coroners would transform a cell or ward into a makeshift morgue. When a corpse was collected by constables from the muddy depths of the Yarra River, coroners would store it in an outbuilding until an inquisitorial hearing could be held. The footprints of coroners determined the itineraries of the dead. In walking through the city, in the performance of their civic roles, coroners not only walked with the dead. They also wrote histories of the dead and collected their memories. In ambulating through alleyways and strolling along promenades, in the routes they took and the trajectories they followed, coroners gathered material for writing an institutional history of the dead.

This chapter traces a spatial history of the death investigation process in Australia in the nineteenth century. It offers a historical account of how the movements of the coroner incorporated the dead into the institutional life of coronial law. The proximity of the dead to the living emerged as a spatial problem in Australia in the late eighteenth and early nineteenth centuries. The unburied corpse not only posed a danger to the physical health of urban dwellers, but its presence was also believed to breach the boundaries between the realms of the living and the dead. Coroners were appointed by colonial governments to collect, identify and investigate the remains of the dead, but also, importantly, to remove the unburied corpse from the public sphere. However, the manner in which they moved, the way they carried the dead through the streets of the city and how they set up inquests in taverns and stored remains in outhouses revealed the different ways in which they formed legal relations with the dead. The itinerant coroner harnessed a range of jurisdictional technologies, such as walking, hawking and building, to establish a lawful place for the dead in the city.

1 *The Argus* (Melbourne), 11 November 1852, 5.

Of dead places

Transformations in attitudes towards the dead shaped the spatial arrangement of Western cities in the nineteenth century.[2] By the late eighteenth century, cities exhibited a morbid curiosity with the 'death of the other' (*la mort de toi*), which manifested in the personalisation of funerary rituals and a romanticisation of the cult of the dead.[3] Le Cimetière du Père Lachaise, Paris's most opulent resting place for the dead, opened on the outskirts of the city in 1804, and inspired a new wave of construction of garden cemeteries across the West. In 1833, Kensal Green, which was designed by John Claudius Loudon as a British counterpoint to Père Lachaise, was established as the first of many botanical necropolises in London. It was followed by Norwood in 1837, Highgate in 1838, Abney Park and Nunhead in 1840 and Tower Hamlets in 1841. As Catharine Arnold puts it, by the 1850s 'London was more necropolis than metropolis, her bustling highways paved with good for the fortunate few, her side-streets reeking of decay [for everyone else]'.[4]

Western cities were transfigured by a rapid construction of elaborate grave-yards, baroque mausoleums and moribund alleyway attractions in the nineteenth century. The dead were increasingly buried in individual tombs with personal-ised inscriptions, routinely visited by survivors seeking some sort of reassurance of their afterlife. Meanwhile the urban-dweller embraced moribund titillations in arcades and laneways, ranging from penny dreadfuls and macabre theatre, to waxworks and the scaffold. Even the Parisian morgue became a tourist attrac-tion, exhibiting daily the wretched seas of human mortality. Signs of death were everywhere in the burgeoning metropolis. Yet, at the same time, the presence of the dead was something to be feared.[5] The popular acceptance of miasmatic the-ories of disease causation depicted the human cadaver as a physical, moral and telluric threat to all living beings.[6] By the end of the nineteenth century, the

2 Philippe Ariès, *The Hour of Our Death: The Classic History of Western Attitudes Toward Death Over the Last One Thousand Years* (Helen Weaver trans, Vintage Books, 2nd edn, 2008). On a social history of the dead and the city, see further Michel Ragon, *The Space of Death: A Study of Funerary Architecture, Decoration, and Urbanism* (Alan Sheridan trans, University Press of Virginia, 1983); Thomas W. Laqueur, *The Work of the Dead: A Cultural History of Mortal Remains* (Princeton University Press, 2015).

3 Philippe Ariès, *Western Attitudes toward Death: From the Middle Ages to the Present* (Patricia M. Ranum trans, The Johns Hopkins University Press, 1974) 56.

4 Catharine Arnold, *Necropolis: London and its Dead* (Pocket Books, 2007) 94.

5 Sigmund Freud offers a psychoanalytic explanation of this ambivalent attitude towards the dead in 'Thoughts for the Times on War and Death' (1915) in *On the History of the Psycho-Analytic Move-ment, Papers on Metapsychology and Other Works, Volume XIV (1914–1916)* (James Strachey trans, Vintage, The Hogarth Press and the Institute of Psycho-analysis, [1915] 2001 edn).

6 For an extended analysis of the reception of miasma theory in Australia, see Marc Trabsky, 'Institutionalising the Public Abattoir in Nineteenth Century Colonial Society' (2014) 40(2) *Australian Feminist Law Journal* 169. On circulation and respiration in the West in the nine-teenth century, see further Alain Corbin, *The Foul and the Fragrant: Odor and the French Social Imagination* (Miriam L. Kochan trans, Harvard University Press, 1986); Richard

places of the dead – from the unburied corpse to the prison cemetery to the garden necropolis – were to be respected and dreaded, venerated and separated from the domains of the living.

The places of the dead *returned* as spatial problems in Western cities in the nineteenth century alongside transformations in attitudes towards the dead. The flesh and the soul reunited in the body of the deceased, a sign of a decline in religious beliefs of the resurrection of the immortal spirit. The living demanded that the mortal body be housed, buried and allotted a singular place within the city. As Michel Foucault writes in 'Of Other Spaces':

> from the moment when people are no longer sure that they have a soul or that the body will regain life, it is perhaps necessary to give much more attention to the dead body, which is ultimately the only trace of our existence in the world and in language. ... it is from the beginning of the nineteenth century that everyone has a right to her or his own little box for her or his own little personal decay.[7]

Cemeteries were subjected to a different kind of spatial arrangement within the city. Not only were they progressively moved from inner-city churchyards to the outskirts of suburbs, but they appeared as counter-sites to the domains of the living, insofar as they inverted, contested and reversed the images of a street, garden and park, while remaining firmly entrenched in those representations. The necropolis replicated an obverse residential estate with its arrangement of separate dwellings for the dead: mausoleums, crypts and charnel houses. It also conjured a paradoxical space: an absent space where time stood still, yet a place for regeneration that was perpetually in motion. The time of the cemetery was incongruous and asynchronous – it marked the end of duration as well as the continuity of life. For Foucault, this *other* place for the dead is 'a place without a place', or what he terms 'heterotopia'.[8]

The concept of heterotopia was introduced to anglophone readers in the Preface to *The Order of Things* in 1970, but was discussed in more detail in a lecture Foucault delivered to architects in 1967, which remained unpublished in English until 1986. 'Heterotopia' was first used in medical discourses to define 'displaced or dystopic tissue ... a [biological] phenomenon occurring in an unusual place'.[9] It was deployed to describe tissue that is out of place, but does not affect the normal functioning of the body. Foucault appropriates this

---

Sennett, *Flesh and Stone: The Body and the City in Western Civilisation* (W.W. Norton & Company, 1996).

7 Michel Foucault, 'Of Other Spaces' (Jay Miskowiec trans, 1986) 16(1) *Diacritics* 22, 25.

8 Ibid, 27.

9 Heidi Sohn, 'Heterotopia: Anamnesis of a Medical Term', in Michiel Dehaene and Lieven De Cauter (eds), *Heterotopia and the City: Public Space in a Postcivil Society* (Routledge, 2008) 41. See also Kevin Hetherington, *The Badlands of Modernity: Heterotopia and Social Ordering* (Routledge, 1997) 42.

anatomical term to denote institutions and places that are anomalous, because they 'neutralize, or invert the set of relations that they happen to designate, mirror, or reflect'.[10] He imagines the cemetery as an exemplary heterotopia, where *another* city is enfolded within the city, an underworld embedded on the surface of the world, a counterpart that interrupts the continuity of the space and time of the living. But this discontinuity is never conceived of as a radical break from the order of things, for the otherness of the cemetery is only experienced in relation to its surrounding milieu.

Unburied corpses threaten to disrupt this constitutive relationship between the city of the living and the heterotopia of the necropolis. They pose a metaphysical danger, as Robert Pogue Harrison points out, to the way humans give meaning to the world. In *The Dominion of the Dead*, Harrison argues that burial practices are integral to 'the humic foundations of our life worlds'.[11] What he means by this is that the act of burying the dead prepares the land for human habitation and makes possible the formation of a place. He writes that burial practices humanise the earth by making sense of land as the groundwork of human history. History can only unfurl through the preservation of the past: 'humans bury not simply to achieve closure and effect a separation from the dead but also and above all to humanize the ground on which they build their worlds and found their histories'.[12]

Burial practices thus transform space into place, which, as Paul Carter notes, is 'a space with a history'.[13] The most obvious way that they do so is through the sign of the grave, which signifies the presence of human place-making. The mark of 'here lies' historicises the ground, removes it from nature and founds a city, a nation on the entombment of ancestors. It 'effectively opens up the place of the "here", giving it the human foundation without which there would be no places in nature'.[14] Thus, when the dead lay unburied in the streets of the nineteenth-century city, the heterotopia of the cemetery threatened to spread throughout the *topos* of the city. The places of the dead were feared to be at once everywhere and nowhere, and the city, to cite again Arnold's description of London, was believed to resemble more a necropolis than a metropolis.

This spatial problem of the unburied corpse was evident in the decision to appoint a coroner for the periphery of the Colony of New South Wales in the 1840s. Charles La Trobe, the superintendent for the then district of Port Phillip, which later became the Colony of Victoria, appointed Dr William Byam Wilmot to the role in 1841. Previously, Captain William Lonsdale, the resident magistrate of the district, fulfilled the duties of the office, alongside a rudimentary group of

10 Foucault, above n 7, 24.
11 Robert Pogue Harrison, *The Dominion of the Dead* (The University of Chicago Press, 2003) x.
12 Ibid, xi. Harrison also writes that 'humanity is not a species ... it is a way of being mortal and relating to the dead. To be human means above all to bury'.
13 Paul Carter, *The Road to Botany Bay: An Exploration of Landscape and History* (University of Minnesota Press, 2010) xxiv.
14 Harrison, above n 11, 20.

police magistrates and justices of the peace. Duties included investigating the identity of the dead and the causes of death, where a person died suddenly, violently, accidentally, or while detained in a penal institution. The historical records of the Melbourne Court Register reveal that, from 1836 to 1840, coronial inquests did not formally take place in the district.[15] The dead were often buried by colonists without waiting for a medical practitioner to conduct post-mortem examinations, while Lonsdale and his clerks merely recorded witness depositions to sudden or violent deaths.[16] Whether based on Christian precepts on the sanctity of the soul, miasmatic theories of disease causation, or simply fear that the unburied corpse would become prey to ravenous animals, witnesses justified burying the dead as 'the best method that could be followed for preservation until the necessary legal steps could be taken'.[17] But such rituals frustrated the death investigation process, obfuscated the conduct of autopsies and compromised a thorough analysis of death scenes. They disrupted the efficacy of the death investigation to the extent that Lonsdale sought advice as to whether 'the sum of five shillings [be] allowed in each case, for finding a dead body subject to a Coroner's inquest'.[18] It was presumed that only a monetary reward could transform the attitudes of the living towards the proximity of the unburied corpse.

La Trobe's decision to appoint a coroner for the district of Port Phillip, who would be tasked to collect, identify and examine human remains in the case of violent and sudden deaths, could be seen as an integral and more than merely incidental part of the British imperial project of colonising the land. Melbourne was a contested place in the mid-nineteenth century. The Indigenous population that occupied the land prior to the British invasion was 'posited as antithetical to the creation and sustainability of an ordered and orderly social space through

---

15 'Chapter 21: Deaths and Inquests', in Pauline Jones (ed), *Historical Records of Victoria – Volume One: Beginnings of Permanent Government* (Melbourne University Press, 1981). Yet, according to several anthropological studies, traditional inquiries into death took place in Aboriginal societies prior to the dispossession of their lands by British settlers: Law Reform Commission of Western Australia, *Aboriginal Customary Laws*, Project 94, Discussion Paper (2005) 300. See also Raymond Brazil, 'Respecting the Dead, Protecting the Living' (2008) 12(SE2) *Australian Indigenous Law Review* 45.

16 Public Records Office of Victoria, *Agency VA 2263 Coroners Courts: History of Coroner's Courts* (2005) Public Records Office of Victoria. www.access.prov.vic.gov.au/public/compo nent/daPublicBaseContainer?component=daViewAgency&breadcrumbPath=Home/Access %20the%20Collection/Browse%20The%20Collection/Agency%20Details&entityId=2263

17 'John Buffington, Alias Ramsay, Shot on Manifold's Station, 3 February 1837', in Jones, above n 15, 307. This was also evident after the appointment of a coroner for the Colony of South Australia in 1839. In one inquest, the coroner declared that 'the body was in a decomposed state and represented a health hazard and this was considered sufficient justification for burial prior to the inquest': Derrick J. Pounder, 'Death Investigation in Early Colonial South Australia, 1839–40' (1984) 24(4) *Medicine, Science and the Law* 273, 277. But see *R v Clerk* (1702) 2 Salk 277; 91 ER 328, where it was held that 'to bury the body before, or without sending for the coroner, is a misdemeanour'.

18 'Reward of Five Shillings for Finding Bodies Subject to Inquest, 26 June 1838', in Jones, above n 15, 310.

which the settlement, the colony and the Empire invented and inhabits a place'.[19] The imperial project of inhabiting a place by burying white settlers on Indigenous lands involved forcefully reiterating the fictional doctrine of *terra nullius* – the notion that the land lacked a history of human habitation. Burial rituals formed part of a suite of technologies, which included enclosing, surveying and mapping the land, through which the British Empire could inhabit, occupy and emplace Indigenous lands. In this sense, the practice of burying the dead prior to the arrival of the coroner can be seen as an attempt to eradicate or remove the legacy of the Indigenous people and their ancestors who already dwelled within those lands. For, if burial practices humanise the earth, transform space into a place and condition the possibility of human history, the actions of colonists were nothing other than a violent attempt to erase and rewrite the history of those lands. While coroners were never obliged to bury the dead – in some circumstances they were even required to exhume the dead – by walking with them through the city, they also participated in this imperial project, albeit in different ways. Coroners wrote a colonial history by assuming responsibility for the movements of the unburied corpse through the streets of the city.

In his request for the appointment of a coroner, Lonsdale cited the demographic growth occurring in the late 1830s and a concomitant increase in the number of unidentified corpses appearing on the streets and in the rivers flowing through the district. His letter to La Trobe also expressed a concern that 'the public would be better satisfied if these inquiries [into the causes of death] were made in the accustomed manner before a Coroner and jury than before a Magistrate only'.[20] The absence of coronial inquests in the district created the impression that the office of coroner, insofar as it was occupied by the resident magistrate, was derelict in its duties towards the dead. However, the public's dissatisfaction with the conduct of death investigations had more to do with the appointment of 'Military men to Civil duties' than with any dereliction of responsibility.[21] The appointment of a civilian to the role of coroner was intended to restore public confidence in the way that the colonial government addressed the spatial problem of the proximity between the living and the dead.

19 Kathryn Ferguson, 'Imagining Early Melbourne' (2004) 1(1) *Postcolonial Text* 1, 1. On a history of coronial investigations into Aboriginal deaths in Australian colonies, see Mark Finnane and Jonathan Richards, '"You'll Get Nothing Out of It"? The Inquest, Police and Aboriginal Deaths in Colonial Queensland' (2004) 35(123) *Historical Studies* 84; Amanda Nettelbeck and Robert Foster, 'Colonial Judiciaries, Aboriginal Protection and South Australia's Policy of Punishing "with Exemplary Severity"' (2010) 41(3) *Australian Historical Studies* 319.

20 'Increase in Sudden Deaths Requires a Coroner, 31 January 1840', in Jones, above n 15, 311.

21 'Port Phillip Gazette, 23 January 1841' quoted in Jim Reid, *The Life of Dr William Byam Wilmot M.D. (1805–1874)* (Unpublished, Victorian Institute of Forensic Medicine, 2001) 12. See further for an informative analysis of the life of Wilmot, Stephen Cordner and Fiona Leahy, 'Forensic Medicine and the Supreme Court', in Simon Smith (ed), *Judging for the People: A Social History of the Supreme Court in Victoria 1841–2016* (Allen & Unwin, 2016) 235–260.

In the 1840s and 1850s, the jurisdiction of the coroner extended to paupers lying face-down in muddy streets, intoxicated cadavers prostrate in rowdy taverns, drowned migrants floating in rivers and murdered remains hidden in private lodgings. The way coroners collected the dead from where they lay, walked with them through the city, carried them from one public house to another and stored them in outhouses when an inquest was delayed reveals much about how they transformed spatial relations with the dead.

The itinerant coroner

Court sittings were peripatetic in Australia in the late eighteenth and early nineteenth centuries. Judicial proceedings took place wherever a magistrate could find temporary accommodation, which included public houses, hotels, churches, schools and hospitals. The first courthouse in Melbourne consisted of 'a wattle and daub hut'; however, the 'police magistrate … moved around at will, setting up shop wherever whim or duty dictated'.[22] By the mid-nineteenth century, court sittings had moved from temporary multipurpose buildings into specially designed courthouses. The history of the movements of coroners largely followed this trajectory. Prior to the construction of purpose-built courthouses in the 1880s, coroners would travel to wherever the dead lay and hold inquests at the nearest available public house or hotel.

Moving across the English countryside to 'keep the pleas of the Crown', medieval coroners unravelled a narrative of itinerancy. The history of the office of coroner in England depicts an itinerant who, from the twelfth century onwards, travelled across the country collecting forfeitures, amercements and deodands on behalf of the King. In *De Republica Anglorum*, which was published in the sixteenth century, Sir Thomas Smith conveyed a similar image when he wrote '[t]he empanelling of thes enquest, and the viewe of the bodie, and the giving of the verdict, is commonly in the streete in an open place'.[23] While the fiscal responsibilities of the coroner were abolished by the mid-nineteenth century, the office still possessed the character of an itinerant. If the dead lay in a place other than a prison or hospital, coroners would 'hawk'[24] the corpse from one public house to another in the hope that a friendly publican would let a room for holding an inquest, or at least provide an outbuilding for storing the dead body until a hearing could be held. The *Coroners Statute 1865* (Vic), for example, obliged all publicans in the city of Melbourne to accept any request from a coroner or constable to accommodate a corpse on their premises

---

22 Michael Challinger, *Historic Court Houses of Victoria* (Palisade Press, 2001) 17.

23 Sir Thomas Smith, *De Republica Anglorum* (Cambridge University Press, [1563] 1906 edn) 91–92 quoted in Clare Graham, *Ordering Law: The Architectural and Social History of the English Law Court* (Ashgate, 2003) 239.

24 The word 'hawk', which connotes the practices of a commercial traveller, was used by Coroner Wilmot in his correspondence with the Colonial Secretary on 14 January 1853: PROV, VPRS 1189, P0, Unit 128, Item 53/446.

for the purposes of an inquest.[25] If they accepted the corpse, they would be financially compensated, while refusal would result in a steep fine. Only towards the end of the nineteenth century were publicans allowed to refuse entry to a corpse in an offensive state of decomposition.

The role that the public house played in the death investigation process sheds light on how the contingency of place affected the administration of coronial law in Australia. No permanent site was ever reserved in Robert Hoddle's rigid survey of Melbourne in 1837 for the purposes of holding coronial inquests or storing dead bodies.[26] 'The idea of a central morgue', Andrew Brown-May writes, 'where bodies would be kept for identification and inquest, was a new one in the urban culture of the nineteenth-century British Empire'.[27] Nor were there any discussions during meetings of the Melbourne City Council of the problem of accommodating the dead until 1852.[28] Hence, when Wilmot assumed the role of coroner in 1841, Melbourne lacked public buildings or structures of any kind to serve as a central place to inspect, collect, or store corpses during the death investigation process.

In a letter addressed to the Colonial Secretary in 1853, Wilmot wrote with much dismay about the spatial problem of accommodating the dead in the city:

> In the present crowded state of the City, the danger attending to the introduction of bodies perhaps in an advanced stage of decomposition into public houses for the purpose of an inquest must be obvious, and the disgraceful scene which took place on Sunday evening last when a corpse was hawked about the streets before any publican would admit it upon his premises, induces me to urge this matter upon His Excellency. There is great allowance to be made for some publicans in this matter, who have had premises licensed without the compliment of stabling and out offices prescribed by the law.[29]

25 Section 11 of the *Coroners Statute 1865* (Vic) set out the legal requirement that '[e]very holder of a publican's license', if required by the coroner or constable, must 'receive into [his or her] house … any dead body that may be brought to such house for the purpose of an inquest being held thereon'. The section was condemned by members of the licensing board who objected to the obligation on a number of grounds: *The Argus* (Melbourne), 5 May 1858, 6.

26 Andrew Brown-May and Simon Cooke, 'Death, Decency and the Dead-House: The City Morgue in Colonial Melbourne' (2004) 3 *Provenance: The Journal of Public Record Office Victoria* 4, 6. http://prov.vic.gov.au/publications/provenance/provenance2004/death-decency-and-the-dead-house

27 Andrew Brown-May, 'History and Development of the Site', in Andrew Brown-May and Norman Day, *Federation Square* (Hardie Grant Books, 2003) 18.

28 Melbourne City Council Minutes, 7 June 1852, Vol. 6 (1894).

29 Correspondence between Coroner Wilmot and Colonial Secretary, 14 January 1853, PROV, VPRS 1189, P0, Unit 128, Item 53/446. The 'disgraceful scene' alluded to in his letter was summarised by *The Argus* (Melbourne), 11 January 1853. In short, on Sunday, 9 January 1853, Wilmot 'hawked [a corpse] from house to house' until 'it was accepted by the Friend-In-Hand'.

He was clearly enraged by the refusal of publicans to receive a corpse in their lodgings. In fact, Wilmot believed that taverns and their outhouses provided the only practical solution to the problem of finding a place, in a city bereft of a morgue or other adequate facilities, for performing inquests and storing dead bodies.

Historians have suggested several reasons as to why public houses emerged in Australia as the most practical solution to the problem of where to accommodate the dead in the nineteenth century. These reasons are in addition to the fact that holding inquests in public houses had a long history in England and continued to be practised in parts of the country until the early twentieth century.[30] The 1850s witnessed an exponential growth in the population of the Colony of Victoria due to the discovery of gold in the north-western town of Ballarat, which in turn saw an increase in the number of unidentified, mostly migrant corpses appearing in the city of Melbourne. Given the rise in the number of inquests held each year to determine the identity of these corpses, public houses enabled Wilmot, first, to provide the public with an opportunity to inspect and where possible identify the dead and, second, to avoid as much as possible conveying the dead through the 'crowded' streets of the city. While he did not specify in his correspondence with the Colonial Secretary the 'obvious' dangers of hawking the dead from one public house to another, as previously mentioned, the lack of a place to store the dead emerged as a problem in Melbourne in the nineteenth century. As mentioned earlier, the practice of carrying the dead through the streets of the city troubled the porous boundaries between the realms of the living and the dead, and between the metropolis and the necropolis.

Conducting inquests in public houses conveniently provided Wilmot with a steady stream of available, though often intoxicated, jurors, willing to sit in front of a corpse for a small fee.[31] Due to the unbearable heat of the Antipodean summer, sites of

30 There are several differences between how this history of itinerancy unfolded in England. What is most notable is that, while inquests were likewise held in public houses in the nineteenth century, the London County Council, which was created in 1889, 'took over administrative control of London inquests after the 1888 Local Government Act' and mounted an impressive 'campaign against public inquests': Ian Burney, *Bodies of Evidence: Medicine and the Politics of the English* Inquest, *1803–1926* (Johns Hopkins University Press, 2000) 204 fn 2. The gradual transition between pub and court inquests in England can be contrasted with the urgency and immediacy of finding a solution for the spatial problem of the place of the dead in Australia.

31 The requirement that an inquest be held before a jury was enshrined in the *Coroners Statute 1865* (Vic) s 5. The *Coroners' Juries Act 1887* (Vic) s 4 set out a scheme for paying jurors a small fee for attending coronial inquests. However, Section 2 of the *Coroners Act 1903* (Vic), which amended previous legislation, granted coroners the power to hold an inquest without a jury, unless required by another legal institution. It is worth noting that the custom of recruiting jurors in public houses incidentally led to accusations in England and Australia that the coroner's inquest was a farce. See further, Ian Freckleton and David Ranson, *Death Investigation and the Coroner's Inquest* (Oxford University Press, 2006) 49. For a history of the public house in England, see Brian Harrison, *Drink and the Victorians: The Temperance Question in England, 1815–1872* (Keele University Press, 1994).

imbibition were the most conveniently situated to conduct an inquest. The reason for this was that heat accelerated the decomposition process and therefore threatened to compromise evidence of the death scene, and the only way to avoid the horrid experience of holding an inquest before a jury upon view of a partially decomposing body was to hold it as close as possible to the site of death and as soon as possible after death occurred. The alternative option of storing dead bodies in the outhouses or stables of taverns was not ideal, and clearly constituted the subject of complaints from many disgruntled publicans. But this latter option was often necessary given that Coroner Wilmot lived in a quasi-rural seaside town, and often took a number of days to travel to the city to hold an inquest.[32]

Wilmot did not simply complain about the behaviour of publicans in his correspondence with the Colonial Secretary. He also proposed a site for his office and a morgue near the embankment of the Yarra River, which was later described as 'a kind of catacomb, which will be marked with shrubs'.[33] He even sketched designs for a courthouse building that would eventually become the coroner's office – a neat low building with a fence in which to conceal any offence to the public[34] – and he urged the then Lieutenant Governor La Trobe to promptly comply with his requests.[35] Wilmot linked the construction of a proper building with a proper name to the effective performance of his role:

> an increasing need arises for an office for my department, it's [*sic*] duties now absorb my undivided attention and I feel it requisite to be completely identified with the offence in order to seem [*sic*] me from the constant interruptions to which my professional position [require].[36]

Ian Burney suggests that critics of the nineteenth-century inquest in England denounced its location in the public house for its profanation of institutional life

---

32 It must be considered that corpses were sometimes stored for days on end, most likely due to inclement weather delaying the coroner's journey into the city. This led *The Argus* to surmise that 'the ends of justice must often be endangered, if not actually defeated, before the attendance of the coroner can be obtained': *The Argus* (Melbourne), 25 June 1849. For a strident censure of Wilmot's languor, see 'The Lame, the Halt and the Blind', *The Argus* (Melbourne), 12 October 1855.

33 *The Argus* (Melbourne), 13 May 1854, 5.

34 The designs of the building cannot be found in the Public Records Office; however, its description is contained in Correspondence between Coroner Wilmot and Colonial Secretary, 29 March 1853, PROV, VPRS 1189, P0, Unit 128, Item A53/3173.

35 Correspondence between Coroner Wilmot and Colonial Secretary, 2 March 1853, PROV, VPRS 1189, P0, Unit 128, Item A53/2203. The Mayor of Melbourne likewise informed La Trobe of the hardship of licensed owners being required to accommodate bodies awaiting inquests in crowded public houses. He requested the erection of a deadhouse along the Yarra River: Correspondence between Mayor of Melbourne and Colonial Secretary, 23 September 1853, PROV, VPRS 1189, P0, UNIT 128, Item C53/8470.

36 Correspondence between Coroner Wilmot and Colonial Secretary, 2 March 1853, PROV, VPRS 1189, P0, UNIT 128, Item A53/2203.

and its celebration of 'civic popular liberties'[37] – not to mention its characteristic stench of alcohol, tobacco and gossip. Only an inquest conducted in the private sphere, in a proper building with a proper name, could be elevated to 'a more technical, administrative, and jurisdictional plane'.[38] Yet, while 'the pub inquest was disparaged as the exemplary sign of inefficiency, disorder, and irrational archaism', Burney further writes, 'the frustrations to scientific inquiry that it represented were also recognized as springing from important – and in some senses necessary – political exigencies'.[39] Certainly the 'public inquest' was venerated by some as politically indispensable for public confidence in the administration of justice. However, as this chapter will now consider, the place-making activities of coroners, which included walking with the dead, hawking them from one public house to another and holding inquests in crowded taverns, affected the way they formed legal relations with the dead. The itinerant coroner transformed spatial relations between the living and the dead by incorporating the latter into the institutional life of coronial law.

Building a deadhouse

In 'Building, Dwelling, Thinking', Martin Heidegger explores the syntactic origins of the German word for 'building'. He argues that the etymology of building belies an ineluctable relationship with the vocabulary for dwelling. This relationship derives from '[t]he Old English and High German word for building, *buan*, [which] means to dwell. This signifies: to remain, to stay in a place'.[40] The verb *bauen* in modern German means 'to build', which moreover signifies not only the practice of dwelling, but also its manner or modality. 'To be a human being means to be on the earth as a mortal', Heidegger writes, 'It means to dwell'.[41] The notion that building is dwelling, and dwelling is being, is advanced by the philosopher's assertion that cultivating constitutes the modality by which to dwell, the 'means to be on the earth as a mortal'. To cultivate is to produce, to 'bring forth' not only buildings but also being, to nurture the growth of oneself. It is in this sense that Heidegger reveals a syntactic link between the vocabulary for building, dwelling and cultivating. If a modality of dwelling is cultivating, and cultivating is a technique of making 'something appear', then building is the construction not simply of a structure to house beings, but also of the forms by which these beings take care of their selves on the earth.[42]

Following the writings of Heidegger, Harrison suggests that language is the ultimate expression of dwelling. He contends that, before dwelling in place,

37 Burney, above n 30, 81–82.
38 Ibid, 82.
39 Ibid, 83.
40 Martin Heidegger, 'Building, Dwelling, Thinking', in *Poetry, Language, Thought* (Albert Hofstadter trans, Harper and Row Publishers, 1971) 146.
41 Ibid, 147.
42 Ibid, 159.

humans are dwelling in *logos*. *Logos*, as Harrison understands it, is something irreducible to language. While the word undoubtedly denotes language in Ancient Greek, it also signifies *relation*: '*Logos* is that which binds, gathers, or relates. It binds humans to nature in the mode of openness and difference. It is that wherein we dwell and by which we relate ourselves to this or that place'.[43] The ineluctability of language and dwelling is precisely what Carter gestures towards in his historical account of the spatial constitution of Australia in *The Road to Botany Bay*. Place-making is an activity of dwelling in space. But the manner of this dwelling makes use of a multitude of techniques, including walking and story-telling, surveying and mapping, building and naming places. The purpose of naming a place 'was to preserve the means by which [names] came to be known, the occasion of places, the sense in which places are means, not of settling, but of travelling on'.[44] The making of a spatial history, the inhabitation of a place, the discovery of lands, begins first and foremost with naming and walking. In other words, both language and movement constituted techniques of place-making by which settlers travelled across, dwelled within, built upon and cultivated relations with Indigenous lands. They were means through which history could congeal into a place. Place-making was not simply a technique of appropriating lands, but also a means of transforming space into a place and inscribing within it an imperial history, in this case a history of the movements of the British Empire.

Coroner Wilmot's insistence upon a proper place and proper name for a deadhouse was concomitant with techniques that had as their aim the 'bringing forth' of a place for the dead. This was a place of dwelling, where not only could the dead appear, gather and remain, but also where the coroner could bind the death investigation process to a lawful place. In walking with the dead through the streets of the city, the coroner transformed spatial relations between the living and the dead. But by petitioning the colonial government to construct a proper place to house the dead, and even offering designs for the building of a coroner's office, Wilmot went much further than this. His actions revealed that these spatial relations first and foremost belonged to law. They showed that walking, hawking and building, as place-making activities of the coroner, attached the dead to a place and that place to the conduct of coronial law. The places through which he walked, the sites where he conducted inquests and the houses where he stored the dead were exposed by Wilmot's designs as integral to the way he performed the duties of his office.

It could be said that this is nothing new. Legal historians have long argued that the effective performance of legal institutions in Australia was contingent upon the construction of specifically designed buildings to house court

43 Robert Pogue Harrison, *Forests: The Shadow of Civilization* (The University of Chicago Press, 1992) 200.
44 Carter, above n 13, 32.

sittings.[45] In Melbourne, though, the absence of a coroner's courthouse did not so much render the coronial inquest ineffectual – as Burney notes of similar practices in London, many found the informalities of the public inquest 'admirable'.[46] Rather, the absence of a purpose-built site for the conduct of death investigations revealed that the techniques of walking, hawking and building were integral to the performance of the office of coroner. When the coroner walked with the dead through the streets of the city, when Wilmot conducted inquests in taverns and hotels, the world of the living collided head on with the places of the dead. It was precisely this proximity between the living and the dead that obliged the coroner to try to make possible a lawful place for the dead in the city.

The inaugural Coroner of Melbourne was only partially successful in petitioning La Trobe for the construction of a proper place for his office and to house the dead. While everyone agreed with Wilmot that '[d]ead bodies in varying states of decomposition should be seen as little as possible', La Trobe and the Colonial Architect disagreed with Wilmot on the proposed location of his office.[47] La Trobe found the mooted site 'horrible'.[48] Despite such criticisms, Wilmot's office was built in 1854 on the corner of a main thoroughfare, which *The Argus* chided as infringing upon the 'busy street-life of a bustling city', while a morgue was added to the site in 1871.[49] The subsequent city coroner, Dr Richard Youl, was more successful in lobbying the government for a purpose-built structure for conducting inquests and storing dead bodies in the city. His tenure oversaw the passing of the *Melbourne Morgue Site Act 1886*

---

45 For a historical analysis of court buildings in Australia, see J.M. Bennett, 'The Evolution of Court Houses in New South Wales', in Terry Naughton, *Places of Judgment: New South Wales* (Law Book Co, 1987); Russell Hogg, 'Law's Other Spaces' (2002) 6 *Law Text Culture* 29.

46 Burney quotes here Joshua Toulmin Smith: 'It brings the eye of the Law directly home to every spot; so that no fact shall escape the searchingness of direct inquiry, and that every man shall know and feel the immediate presence of the course of Justice': Burney, above n 30, 206 fn 11.

47 Correspondence between Coroner Wilmot and Colonial Secretary, 2 March 1853, PROV, VPRS 1189, P0, UNIT 128, Item A53/2204. It should be noted that the initial plans for the office included a room for the coroner; another for the registrar of births, deaths and marriages; and a post-mortem examination room. See Correspondence between Colonial Engineer and Colonial Secretary, 26 May 1854, PROV, VPRS 1189, P0, UNIT 128, Item E54/5695; Correspondence between Auditor General and Colonial Engineer, 18 December 1854, PROV, VPRS 1189, P0, UNIT 128, Item 54/14059.

48 Correspondence from Lieutenant General, 26 April 1854, PROV, VPRS 1189, P0, UNIT 128, Item E54/5695.

49 *The Argus* (Melbourne), 25 August 1886, 4. The architectural blueprint of both buildings depicted a separate room for post-mortems and another for inquests. It set out an office for the coroner and a yard separating the morgue from the courtroom. Both buildings were compulsorily acquired by the Railway Department in 1883. The Melbourne Fish Market stood opposite the coroner's office from 1865 to 1892. It was demolished in 1900 to make way for the new Flinders Street train station. The location of the office can be gleaned from railway maps from the late nineteenth century: see, e.g., Victorian Railways, Melbourne and Oakleigh Line, Plan of Land, PROV, VPRS 266, Unit 413, Item 87/3691.

(Vic) and the construction of a specially designed coroner's courthouse along the Yarra River in 1888.[50] Though, as Alfred Deakin announced in Parliament in 1886, even the proposed location of the new coroner's courthouse was problematic:

> The [city] corporation objected to any site which they considered would be disfigured by the erection of a morgue, which, on account of its association, was supposed – however perfect it might be from an architectural point of view – to be an undesirable object to be placed in any public position.[51]

Persistent calls for the construction of a deadhouse in the city were not simply idiosyncrasies of piqued coroners. Such concerns were equally shared by jurors, councillors, undertakers, journalists and other legal officers of the colony. When Youl, as acting city coroner, sent a report in 1855 to a committee investigating 'the condition and accommodation of Government Offices', and implored the Surveyor General and Colonial Engineer of 'the absolute necessity which exists for the erection of a morgue, in connection with the office of the coroner', he was not merely speaking on behalf of a disgruntled low-level government officer.[52] He was voicing the chagrin of the public who were compelled by the absence of a deadhouse to share their lodgings with the dead. Particularly in the scorching summer days of January, when decomposing corpses lay in public houses awaiting coronial inquests, which, as one journalist intimated, was 'an intolerable nuisance, at this season of the year especially', the plea for a deadhouse could be heard alike from the coroner, publican and

50 The courthouse was designed by the government architect, Mr G.W. Watson of the Public Works Department. Many aspects of the construction process, particularly the costs involved, were criticised by lawyers, councillors, politicians and journalists. The project was substantially delayed for a number of years, even though Crown land was reserved for the courthouse under legislation. On the other hand, the elaborate design was championed by the secretary of the Crown Law Department: 'The Proposed New Morgue', *The Argus* (Melbourne), 19 May 1887, 6. For further analysis of the design of the courthouse, see Marc Trabsky, 'The Custodian of Memories: Coronial Architecture in Nineteenth Century Melbourne' (2015) 24(2) *Griffith Law Review* 199.

51 'Parliament: Legislative Council', *The Argus* (Melbourne), 25 August 1886, 4 and 9. It should be noted that Youl requested a central site that would be convenient for collecting the dead but also summoning a jury: '[T]he Lands office considers its south bank sites too valuable for the purpose; the City Council objects to a prominent Flinders-street situation; and last of all, whatever the locality, aggrieved individuals join in a chorus of indignation'. In addition to this, the Public Works department felt that locating the morgue under Princes Bridge 'would spoil the bridge'.

52 *The Argus* (Melbourne), 27 January 1855, 6. The medically trained Youl was appointed the district coroner for Bourke in 1853. His jurisdiction included the surrounding suburbs of Melbourne, but not the city itself. He was then appointed the acting coroner of Melbourne in 1854. He succeeded Wilmot as city coroner in 1857: Ann Mitchell, *Youl, Richard (1821–1897)* (1976) Australian Dictionary of Biography, National Centre of Biography, Australian National University. http://adb.anu.edu.au/biography/youl-richard-4900/text8201

drunkard.[53] The deplorable situation continued throughout much of the nineteenth century and perhaps came to a head largely as the result of the Windsor Railway accident in 1887, which not only rendered visible to the public the problem of accommodating the dead in the city, but even stymied the conduct of the death investigation process: 'So bad was the stench, the filth and the whole of the surroundings, that the jury would not proceed with the business, and after viewing the bodies, the inquest had to be adjourned to a suitable building'.[54]

Walking in the necropolis

The practice of walking with the dead through the streets of the city in the nineteenth century unsettled the porous boundaries between the city of the living and the heterotopia of the necropolis. It also transformed the spatial relations between the living and the dead, albeit in a different way from the building of a deadhouse. In *The Practice of Everyday Life*, Michel de Certeau identifies the activity of walking through the city as a unique mode of operating, one that is different from the activity of dwelling. Walking is a practice of storytelling, which disrupts the rational layout of the city and 'repress[es] all the physical, mental and political pollutions that would compromise it'.[55] On the one hand, to walk is to be without a place, to be lost in space; yet, as a rhetorical style, it is a speech act that enunciates place. Place is determined by proprietary – a proper name, mark or sign – whereas space is characterised by the unsettling action of movement. The corpse is, for de Certeau, a paradigmatic symbol of place: 'the *being-there* of something dead, the law of a "place"'.[56] To cite Heidegger, the corpse is dwelling par excellence. In comparison to the inertness of the corpse, walking is a dynamic technique of place-making, a craft that 'constantly transforms places into spaces or spaces into places'.[57] It is a practice that cultivates spatial relations with proper names, objects and persons.

53 *The Argus* (Melbourne), 3 January 1855, 5. On 3 January 1855, Coroner Youl, believing that the erection of the morgue was imminent, arranged with Mr Crofts, the undertaker, to store dead bodies found in the city at his premises in Queensberry Street, North Melbourne. Each corpse would be stored in the outbuilding until an inquest could be held. It seems that the scorching temperatures during December precipitated this informal arrangement, as evident by a report in *The Argus* that Alexander McQueen, who drowned in the Yarra River, was 'deposited in a fowl-house' while awaiting an inquest, and in turn, his decomposing body was 'exposed to the heat of the atmosphere', which caused great distress not only for his friends, but also for the journalist: *The Argus* (Melbourne), 1 December 1854.
54 Secretary Harriman to Minister of Justice, Crown Law Office, Memo, 8 July 1887, PROV, VPRS 266, P0, Unit 413, Item 87/3721.
55 Michel de Certeau, *The Practice of Everyday Life* (Steven Rendall trans, University of California Press, 1984) 94.
56 Ibid, 118 (emphasis in original).
57 Ibid.

Walking with the dead is not just any technique though; more than any-thing else, it is a jurisdictional technology that shapes relations belonging to law. Jurisdiction has recently been conceptualised in critical legal scholarship as irreducible to the exercise of sovereignty over territory.[58] This theoretical turn has been inspired by the etymology of the Latin words for 'law' (*ius*) and 'to speak' (*dicere*), which denotes 'the authority to speak the law – an authority presupposing a setting apart of the legal from the non-legal'.[59] In other words, jurisdiction signifies the voice of law, law's speech and the act of speaking law. In looking beyond jurisdiction as the exercise of sovereignty over ter-ritory, Shaunnagh Dorsett and Shaun McVeigh conceive of it as a technical and material practice capable of producing different effects in law. Jurisdiction consists of an inventory of techniques that are irreducible to language, a repertoire of devices that attach themselves to institutional practices and 'mark the existence of legal institutions and give shape to lawful relations'.[60] In *Jurisdiction*, Dorsett and McVeigh present writing, mapping, taxonomy and precedent as examples of the 'technical means by which a conduct of lawful relations is given shape'.[61]

By 'lawful', Dorsett and McVeigh are referring to relations that properly belong to law; in other words, they are describing how technical, material and spatial relations become attached to legal institutions. This chapter, however, employs the concept of lawfulness to describe not only how the dead were attached to the institutional life of coronial law, but also how they came to be bound to a proper place with a proper name. It is in this sense that it is pos-sible to conceive of coroners as occupying the role of Marcel Mauss's techni-cians, who, with the assistance of a wide range of technologies, were capable of attaching to legal institutions spatial relations between the dead, the living and the city.[62]

---

58 On this recent turn in jurisdictional thinking, see Edward Mussawir, *Jurisdiction in Deleuze: The Expression and Representation of Law* (Routledge, 2011); Shaunnagh Dorsett and Shaun McVeigh, *Jurisdiction* (Routledge, 2012); Daniel Matthews, 'From Jurisdiction to Juriswrit-ing: At the Expressive Limits of the Law' (2014) 1 *Law, Culture and the Humanities* 21; James E.K. Parker, *Acoustic Jurisprudence: Listening to the Trial of Simon Bikindi* (Oxford University Press, 2015); Mariana Valverde, *Chronotopes of Law: Jurisdiction, Scale, and Gov-ernance* (Routledge, 2015); Olivia Barr, *A Jurisprudence of Movement: Common Law, Walk-ing, Unsettling Place* (Routledge, 2016). For an earlier analysis of territorial jurisdiction, see Richard T. Ford, 'Law's Territory: A History of Jurisdiction' (1999) 97(4) *Michigan Law Review* 843.
59 Maria Drakopoulou, 'Of the Founding of Law's Jurisdiction and the Politics of Sexual Differ-ence: The Case of Roman Law', in Shaun McVeigh (ed), *Jurisprudence of Jurisdiction* (Rou-tledge, 2007) 33.
60 Dorsett and McVeigh, above n 58, 34.
61 Shaun McVeigh, 'Law as (More or Less) Itself: On Some Not Very Reflective Elements of Law' (2014) 4 *University of California Irvine Law Review* 471, 473.
62 Marcel Mauss, 'Technology (1935/1947)' (Dominique Lussier trans) in *Techniques, Technol-ogy and Civilisation* (Durkheim Press/Berghahn Books, 2006).

In *A Jurisprudence of Movement*, Olivia Barr examines the spatial movements of a burial party in the Colony of New South Wales in the eighteenth century and of a New Zealand coroner in Antarctica in the twenty-first century as jurisdictional technologies that bind the common law to a place. In both examples, the rhythms of the common law, its trajectories and itineraries are entangled with the wanderings of the burial party and the proposed flight lines of the coronial jurisdiction. Burial practices, according to Barr, along with camping and walking, are inextricable not only from how the common law is bound to a place, but also from how the office of jurist (and, to a lesser extent, the office of coroner) forms legal relations with the dead.

This idea is built upon the argument that common law only 'comes into its relations and comes to be in place' through the technology of movement.[63] Or, as Barr puts it, '[w]hile there are many movements, and many meanings, most fundamentally this is movement as the institution and form of common law, and one that is given shape by jurisdiction'.[64] In other words, movement is inherently a jurisdictional technology, and because the institution of burial is a *paradigm* of movement, if not an essential component of how common law moves, burial practices institute a lawful place for the common law. Barr concludes that the way common law moves, the making of its place and the conduct of legal relations are all tied to the burial of the dead:

> For it is through movement that common law comes into its relations with the dead; through movements in ceremonies of burial common law comes to be or at least seem to be in place; and through movement common law creates and conducts lawful relations with the dead.[65]

Barr is inspired here by Harrison's claim that 'places are not only founded but also appropriated by burial of the dead'.[66] For Harrison, the institution of burial is essential for humanity, it is foundational to the building of a house, the making of a history and the humanisation of the earth. It institutionalises a meeting place between the living and the dead, a public sphere where the people that inhabit the ground and the ancestors that dwell within it may gather.[67] Yet, more than this, Barr suggests that burial practices attach common

---

63 Barr, above n 58, 65.
64 Ibid.
65 Ibid, 102. On the one hand, Barr suggests that it is the ceremony of burial itself that institutes 'lawful relations between common law and the dead'. At other times, 'it is not burial per se, but rather movements towards burial that become significant'. This is most apparent in the story narrated in Chapter 3, for 'the series of burials in and around Hawkesbury in 1799 are burials that never quite happen': Ibid, 149–150.
66 Harrison, above n 11, 24.
67 The site where the 1854 coroner's office once stood in Melbourne is called Federation Square. Carter describes it as a meeting place of becoming a 'federal community' of the past

law to the land. She writes that the activities of burial 'are one of the ways in which common law comes to be in place'.[68]

Walking is one of the jurisdictional technologies utilised by coroners in the nineteenth century to form legal relations with the dead. That is to say, coroners attached the dead to law, and law to a place, precisely by *not* burying the dead or moving towards their burial, but by *carrying them towards* the site of an inquest. In walking with the dead through the streets of the city, hawking them from one public house to another – in short, through all of the different place-making activities that have been catalogued in this chapter – coroners bound the dead to legal institutions. In lobbying the colonial government for the construction of a deadhouse, even offering to participate in the design process, coroners once again sought to transform spatial relations between the living and the dead. Through the construction of a purpose-built courthouse to remove the dead from the public sphere, coroners cultivated a lawful place for the dead in the city. Of course, institutions of burial were another mode of forming legal relations with the dead in the nineteenth century, but, as I emphasised throughout this chapter, they did not define the way coroners walked with the dead.

Hauntings of a courthouse

The purpose-built coroner's courthouse in the city of Melbourne was occasionally transformed into a site of conviviality towards the end of the nineteenth century. On 25 May 1897, the inquest room was transfigured, with the careful arrangement of bouquets of flowers and bottles of champagne, into what *The Argus* called a spectacle of 'unusual gaiety'.[69] The festive scene witnessed the unveiling of a photographic portrait of Youl. In honour of his longstanding occupation in the office, and before a gathering of prominent medical and legal professionals, the coroner was presented with a portrait to be hung on the wall of the inquest room. It was later revealed that his portrait was hung directly behind the coroner's chair, implying that, even in his death, 'the old dictator of the Morgue',[70] as he was described in his obituary, would continue to preside over the conduct of the office.[71]

In the following year, the inquest room was once again transformed into a spectacle. On 4 March 1898, a number of pathology staff gathered to witness the unveiling of photographic portraits of the new coroner of Melbourne, Mr Samuel Curtis Candler, and the coroner's surgeon, Dr Crawford Henry Mollison. Both portraits were to be hung in the inquest room 'on each side of the

and the present: *Mythform: The Making of Nearamnew at Federation Square* (Miegunyah Press, 2005) 5.
68 Barr, above n 58, 96.
69 'Presentation to Dr Youl', *The Argus* (Melbourne), 26 May 1897, 6.
70 'Death of Dr Youl', *The Argus* (Melbourne), 7 August 1897, 10.
71 Brown-May and Cooke, above n 26, 4.

photograph of the late Dr Youl'.[72] *The Argus* described the 'convivial gathering' on that day as a scene of 'somewhat incongruous hilarity'.[73] The apparent odd-ness of the event might have been borne of the difficulty of reconciling a jubilant atmosphere with the death-bound images that framed the architecture of the courthouse. The dead were not absent from the courthouse during such gatherings. They were prostrate in the mortuary, albeit hermetically sealed from the inquest room by a thick glass screen. Their presence reverberated through-out the building, for as Jacques Derrida once wrote, the dead never disappear when they pass away. Whether they are memorialised in portraits, effigies, or death masks, or linger in absent objects and empty spaces, the dead always remain at work.[74] Immortalised in their portraits, imprinted on a wall above the coroner's chair, the absent dead surveyed the jubilance of the living.

When Dr James Edward Neild, a renowned forensic pathologist, spoke at the event, he 'warmly *eulogised* both coroners for the manner in which they per-formed their duties'.[75] In 'Death, Decency and the Dead-House', Andrew Brown-May and Simon Cooke interpret this spectacle as evidence of how the office came to understand its place in the administration of colonial government. 'Clearly', they write, 'the morgue was an institution that had a sense of its own past and identity'.[76] However, this event also revealed how the place-making activities of the coroner were imbricated in the formation of legal relations with the dead. The activities of walking, hawking and building furnished for the cor-oner jurisdictional technologies for incorporating the dead into the institutional life of coronial law.

I remarked earlier that the proximity of the unburied corpse in Melbourne unsettled the relationship between the realms of the living and the dead in the nineteenth century. This spatial problem manifested as a fear that the dead were at once everywhere and nowhere, and that they were dwelling in sites of imbibi-tion and hospitality, specifically the tavern and the hotel. The appointment of a coroner for the district of Port Phillip in the 1840s was intended to extricate the dead from the public sphere, remove them from the streets of the living and return them to their *rightful* resting place in the cemetery. It was also supposed to revive the metropolis, in pursuit of an imperial project of colonising the land as the exclusive realm of British settlers. Yet the appointment of a coroner to the district did not exactly achieve those results. Certainly, Wilmot transformed spatial relations between the living and the dead, but the way that he moved, and how he walked and hawked the dead, exacerbated the problem of their proximity in the public sphere. Parading the dead through the streets of the city

72 *The Argus* (Melbourne), 5 March 1898, 9.
73 Ibid.
74 Jacques Derrida, *Specters of Marx: The State of the Debt, the Work of Mourning and the New International* (Peggy Kamuf trans, Routledge Classics, 2006).
75 *The Argus* (Melbourne), 5 March 1898, 9 (emphasis added).
76 Brown-May and Cooke, above n 26, 4–5.

towards the nearest public house, coroners collected a heady mix of intoxicated jurors, hurried reporters and curious onlookers, to participate in an anxiety-ridden, though civic process of communing with the dead.

The construction of the coroner's courthouse in the 1880s was undoubtedly intended, for once and for all, to address the spatial problem of the uncomfortable propinquity between the living and the dead. While it partially succeeded in removing the conduct of inquests from public houses, and constraining the place where the dead lay to a centrally determined and hidden location on the banks of the Yarra River, it cannot be concluded that it simply banished the dead from the world of the living. Rather, the coroner's courthouse granted the dead a lawful place in the city. It was an *institutionalised necropolis*, which framed an architecturally designed encounter between the living and the dead. And it provided an accessible entry point in the city where one could search for missing persons and identify the unknown dead, while also satisfying one's morbid curiosity about the plight of the dead *other*. '[F]lattening [their] noses against the screen of glass', the living could gawk at the most recent 'loads of wretched humanity – the flotsam and jetsam of life's boisterousness'.[77]

The practice of walking with the dead through the city constituted a technology of office in the nineteenth century. The ways in which the coroner travelled to the place where the dead lay, hawked them from one public house to another and conducted inquests in crowded taverns transformed the spatial relations between the living and the dead. Far from expropriating the dead from the public sphere, the jurisdictional movements of the coroner incorporated them into the *polis*. I have suggested in this chapter that writing a spatial history of the movements of the coroner is imperative for accounting for how the unburied corpse came to occupy, and so dwell lawfully, in the city. However, this history also incidentally narrates how the office of coroner became materially placed in the city. Techniques of place-making did not only bind the dead to an institutional life, they attached the conduct of an office to a proper place and proper building.

Coroners were never obliged to bury the dead. Far from it, they were required to carry them, in full view of the living, through the ceremony of coronial procedure. As such, techniques of place-making were an important, if not essential, aspect of the role. While it could be argued that the naming of a proper place to house the conduct of the office mitigated the importance of this obligation, the building of a deadhouse only seemed to reveal, perhaps even more than the conduct of inquests in public houses, how the movements of the coroner remained intrinsic to the performance of office. The idea that the efficacy of the death investigation process depended on activities of walking, hawking and building suggests that the dead became attached to the institutional life of coronial law only through the jurisdictional technology of movement. In walking through the streets of the city, ambulating down alleyways and

---

77 Mannington Caffyn, 'They Met at the Morgue', *The Bulletin* (Melbourne), 9 July 1892, 21.

wandering through its thoroughfares, coroners called for the living to live along-side the dead. They called for the living to learn to share their dwellings with the dead. They even seemed to suggest, as de Certeau writes, that '[h]aunted places are the only ones people can live in'.[78]

78 de Certeau, above n 55, 108.

2 Visual regimes of the dead

The most distinctive feature of the coroner's courthouse built in the late nineteenth century was a hermetically sealed glass screen that separated the mortuary from other parts of the building. The screen effectively segregated courthouses into two zones – one occupied by the dead and another inhabited by the living – and transformed the way in which jurors viewed the corpse during an inquest. Prior to the use of transparent partitions in courthouses, it was commonplace for jurors to view the dead body in the same room as that in which the hearing was held. Only in locations where more space was available would the procedure take place in an adjacent room or even an outbuilding. In purpose-built courthouses, the screen not only enabled jurors to observe the corpse at a distance, but also transformed the sensory experience of the inquest. 'Not only will it be absolutely impossible for any offensive smell to make its way into this passage', *The Argus* newspaper reported at the opening of Melbourne's coroner's courthouse in 1888, 'but [it] has been so successfully arranged that it is as sound proof as a double brick wall'.[1]

The 'air-tight glass passage' instituted an auditory, olfactory and tactile division between the living and the dead. It was sealed such that jurors could not hear the sharpening of instruments or the shuffling of cadavers. They were not able to smell the odours of decomposition or the pungency of formaldehyde. The passage also transformed ocular relations between the living and the dead. It designated the only viewing platform in the courthouse where the living could observe the conduct of post-mortem examinations. The glass screen framed a speculative narrative by translating anatomical wounds into a theatre of intrigue. This frame was not supplementary to the spectacle: it mediated the appearance of the corpse before the inquisitorial forum.

While the use of the glass screen in the Melbourne courthouse was unique in Australia, and heralded its introduction in England, the design was not unprecedented in France.[2] The device was initially developed for the construction of La

---

1 'The New Morgue', *The Argus* (Melbourne), 26 September 1888, 11.
2 Viewing lobbies and inspection windows were also installed in coroner's courthouses in London towards the end of the nineteenth century. However, they were smaller in size from the glass screen installed in Melbourne's courthouse: Clare Graham, 'Sudden Death and the

Morgue in Paris. The practice of displaying cadavers for identification in 'a morgue' originated in France in the eighteenth century; however, the invention of a screen to facilitate such encounters was not utilised in the Parisian morgue until 1864.[3] The use of this device in La Morgue reflected a shift in attitudes towards the dead in France. Death, as Michel Ragon claims, had transformed into 'a spectacle, [it] made a spectacle of itself'.[4] Indeed, the screen projected a sanitised spectacle of the macabre, which, similar to the display cases found in anatomy museums, zoos and aquariums, formalised an institutional relationship between the object of display and the gaze of the observer.

Allan Mitchell provides a vivid description of the Parisian morgue, noting that

> the main room resembled nothing so much as an aquarium. Stretching nearly the length of the hall was a large exhibition case with a panelled glass front, in appearance much like a huge sky-light. Behind the glass façade were two rows of black marble slabs, a dozen in all, illuminated and titled to afford easier viewing of the subjects displayed. Above each headrest was a faucet to permit occasional sprinkling of the corpse with water and chemicals, the only means of preservation employed before the installation of a refrigerating system in the early 1880's.[5]

La Morgue was familiar to the readers of *The Argus*. As early as 1871, the Melbourne newspaper admonished the landmark as 'one of the dismal sights of Paris' and 'a gloomy emanation from the morbid sentimentality of a French mind'.[6] The morgue was decried as '"La Musée de la Mort", a morbid attraction listed in guide books to the city as a must-see alongside the Eiffel Tower, the Louvre, the waxworks and the theatre'.[7] It served multiple purposes in the burgeoning metropolis of Paris – most importantly, as a site for the exhibition of unidentified corpses, collected from the gloomy depths of the Seine River.

LCC: Accommodation for Inquests in London before the First World War' (1995) 1 *Arq* 60, 64–65. The screen was, interestingly, criticised by the *British Medical Journal* as offering an 'unscientific' view of the corpse: 'In all well-equipped mortuaries recently built, the jury actually "view" the body through a window standing outside. How can such a casual glance help the jury to ascertain the cause of death?': '"The View" at Inquests', *British Medical Journal* (London), 1 October 1898, 995.

3 Allan Mitchell, 'The Paris Morgue as a Social Institution in the Nineteenth Century' (1976) 4 *Francia* 581, 581.

4 Michel Ragon, *The Space of Death: A Study of Funerary Architecture, Decoration, and Urbanism* (Alan Sheridan trans, University Press of Virginia, 1983) 137.

5 Mitchell, above n 3, 584. See also *Vue Intérieure de la Morgue en 1845 (Dessin D'après une Peinture de Carré)*, *Gallica Bibliothèque Numérique*, A29479.

6 *The Argus* (Melbourne), 13 January 1871, quoted in Andrew Brown-May and Simon Cooke, 'Death, Decency and the Dead-House: The City Morgue in Colonial Melbourne' (2004) 3 *Provenance: The Journal of Public Record Office Victoria* 4, 8. http://prov.vic.gov.au/publica tions/provenance/provenance2004/death-decency-and-the-dead-house.

7 Ibid, 7–8.

Yet it also served to satiate the morbid curiosity of visitors and tourists who flocked to the morgue in search of the macabre. For Vanessa Schwartz, the morgue was 'a spectacle of the real', a 'grand display' of distressing, titillating and salacious scenes.[8] She quotes an administrative director from the nineteenth century who opined that 'the morgue is considered in Paris like a museum that is much more fascinating than even a wax museum because the people displayed are real flesh and blood'.[9] That the building contained a separate room for photography, the glass windows were curtained to change the display of bodies every three days and most corpses were exhibited nude except for a small cloth covering their genitals only served to heighten the extent to which it appeared as a spectacle of the dead.

The etymology of the word 'morgue' offers insight into how La Morgue came to be associated with transforming ocular relations between the living and the dead. In French, *morgue* designates a place to exhibit corpses for the purposes of identification. Yet as a verb it denotes the practice of identifying the dead. *Morgue* derives from the archaic verb *morguer* – 'to stare, to have a "fixed and questioning gaze" ... to look solemnly'.[10] In the seventeenth century, *morguer* referred to procedures for classifying prisoners upon their entrance to and exit from penal institutions. Guards were required to *morguer* the prisoners upon their arrival at the gaol – that is, to take time to stare at their face and remember their features through a visual inventory. This procedure enabled guards to identify whether a particular individual had died in custody or escaped from prison. In the eighteenth and nineteenth centuries, techniques for inventorying prisoners were introduced into other institutions, including the hospital, factory, workhouse and morgue. It brought to those different institutions new tools for observing the dead.

The hermetically sealed glass screens installed in coroner's courthouses framed visual encounters between the living and the dead. They shone a spotlight on the importance of the post-mortem examination, which remained marginal to the death investigation process until at least the 1860s and 1870s. Prior to this, autopsies were seldom conducted before a coronial inquest. It would suffice for the coroner to merely glance over the surface of the corpse. The development of medical jurisprudence as a sub-discipline of human anatomy transformed the way coroners and jurors viewed the corpse. It positioned the forensic gaze as intrinsic to the legal custom of *super visum corporis*, which meant that an inquest could only take place 'upon view of the dead body'. This chapter offers a historical account of the transformations of techniques for viewing the dead body in the late nineteenth century. It argues that changes to the forensic gaze were integral to how coroners maintained legal relations with the dead. The gaze

8 Vanessa R. Schwartz, *Spectacular Realities: Early Mass Culture in Fin-de-Siècle Paris* (University of California Press, 1998) 48.
9 Ibid.
10 Ibid.

became part of a repertoire of jurisdictional technologies through which the coroner assumed responsibility for taking care of the dead.

Situating the forensic gaze

The science of human anatomy rose to prominence in England at the beginning of the seventeenth century, although it emerged much earlier in continental Europe during the Renaissance. Human dissection was taught at European universities from as early as the fourteenth century: the first recorded public autopsy was conducted by Mondino de Liuzzi at the University of Bologna around 1315.[11] The French word for autopsy, *autopsie*, derives from the Latin *autopsia* and the Greek *autoptes*, which both mean seeing (*optos*) for one's self (*autos*). The restoration of the art of dissection during the Renaissance instituted a new mode of knowing the body. Corporal knowledge became acquired through empirical learning, through direct observation as opposed to hearsay evidence. In the seventeenth century, for example, William Harvey proved by dissecting cadavers that blood circulated throughout the human body, while Giovanni Battista Morgangi developed a hermeneutics of the post-mortem for classifying and interpreting signs of morbid pathology.[12]

Human cadavers were legally supplied to English medical schools for the purposes of pedagogical anatomy from the sixteenth century.[13] However, in the eighteenth and nineteenth centuries, the study of human dissection was marred by controversy.[14] This period witnessed a sharp increase in the number of

---

11 Elizabeth Klaver, 'Introduction', in Elizabeth Klaver (ed), *Images of the Corpse: From the Renaissance to Cyberspace* (University of Wisconsin Press, 2004) xiii. On the history of human anatomy in continental Europe, see Bernard Schultz, *Art and Anatomy in Renaissance Italy* (UMI Research Press, 1985); Jonathan Sawday, *The Body Emblazoned: Dissection and the Human Body in Renaissance Culture* (Routledge, 1995).

12 Ibid, xiv.

13

> In 1540 the Company of Barbers of London united with the Fraternity of Surgeons to form the United Company of Barber-Surgeons [which later became known as the Royal College of Surgeons]. The Charter given [to] them by Henry VIII made provision for the study of anatomy. ... In 1565, Queen Elizabeth I granted the Royal College of Physicians 4 bodies annually for dissection.
>
> Vernon D. Plueckhahn, *Lectures on Forensic Medicine and Pathology*
> (University of Melbourne Printing Services, 5th edn, 1982) 107–108

> It is important to note that *An Act for Better Preventing the Horrid Crime of Murder* (25 George II, c. 37, 1752) made it lawful for prisons to supply medical schools with the bodies of executed criminals in the eighteenth century.

14 The history of human dissection in England, Australia and America has been discussed widely in academic scholarship: Tim Marshall, *Murdering to Dissect: Grave-robbing, Frankenstein and the Anatomy Literature* (Manchester University Press, 1995); Ruth Richardson, *Death, Dissection and the Destitute* (The University of Chicago Press, 2001); Michael Sappol, *A Traffic of Dead Bodies: Anatomy and Embodied Social Identity in Nineteenth-Century*

students studying medicine in London and Edinburgh, which in turn kindled the demand for anatomical cadavers. The proliferation of medical schools in England and Scotland, particularly privately run institutions that were barred from using lawful means to obtain corpses, resulted in the formation of a lucrative underground market for trafficking in dead bodies. The traders in this market, who were also known as the 'resurrectionists', resorted to any means, including body snatching, grave robbing and even murder, to supply cadavers to (private) medical schools.[15] It was not until the enactment of the *Anatomy Act 1832* (UK) that the trade came to an abrupt halt. The Act regulated the study of anatomy in medical schools and established a legal stream for the provision of cadavers by permitting an executor or any party who had lawful possession of a corpse to send it to a medical school in exchange for a fee. The subjects of dissection were still overwhelmingly collected from public institutions, such as workhouses, prisons and hospitals, and were disproportionately represented by convicted criminals, paupers and Indigenous people.[16]

Medical jurisprudence appeared as a specialised sub-discipline of human anatomy in England and Scotland in the nineteenth century.[17] It was initially taught in universities from 1821, but became a compulsory aspect of the medical curriculum by the 1890s.[18] The prominent physician, Samuel Farr, published the

America (Princeton University Press, 2004); Helen MacDonald, *Human Remains: Dissection and Its Histories* (Yale University Press, 2006); Helen MacDonald, *Possessing the Dead: The Artful Science of Anatomy* (Melbourne University Press, 2010); Elizabeth T. Hurren, *Dissecting the Criminal Corpse: Staging Post-Execution Punishment in Early Modern England* (Palgrave Macmillan, 2016); Richard E. Bennett, *Capital Punishment and the Criminal Corpse in Scotland, 1740–1834* (Palgrave Macmillan, 2018).

15 Since the corpse was held under common law to constitute neither a person nor a thing, it was only illegal to steal shrouds, monuments or other items from the grave, but not the cadaver itself. See further, MacDonald, *Possessing the Dead*, above n 14, 7–14.

16 See, e.g., Elisabeth Bronfen, *Over Her Dead Body: Death, Femininity and the Aesthetic* (Manchester University Press, 1992); Richardson, above n 14; MacDonald, *Possessing the Dead*, above n 14; MacDonald, *Human Remains*, above n 14; Cressida Fforde, 'From Edinburgh University to the Ngarrindjeri Nation, South Australia' (2009) 61(1–2) *Museum International* 41.

17 The science of medical jurisprudence originated in China with the 1247 publication of Song Ci's *Washing Away of Wrongs* (*Xiyuan jilu*). See further, Daniel Asen, 'Song Ci (1186–1249), "Father of World Legal Medicine": History, Science, and Forensic Culture in Contemporary China' (2017) 11 *East Asian Science, Technology and Society: An International Journal* 185. The discipline was further developed in the sixteenth century with the *Constitutio Vambergensis Criminalisi* (1507) and *Constitutio Criminalis Carolina* (1532), which advised 'judges to consult surgeons in all cases of suspected homicide, and midwives in suspected infanticides': Catherine Crawford, 'Medicine and the Law', in W.F. Bynum and Roy Porter (eds), *The Companion Encyclopaedia of the History of Medicine* (Routledge, 1993) 1623.

18 Thomas Forbes claims that 'England had lagged literally centuries behind the Continent in the developing field of forensic medicine': 'Crowner's Quest', in (1978) 68(1) *Transactions of the American Philosophical Society* 42. However, Carol Loar argues that forensic evidence and medical expertise had a greater role in the early modern inquest in England than has

first English textbook on the subject in 1788, which was essentially a translation of J.F. Fazelius's *Elementa medicinae forensis* (1767). Farr wrote that

> the examination of the dead body should be as soon as possible after death, in the day time, at a proper place, where a dissection, if necessary, (and it is almost always necessary) may be performed, and not according to vulgar custom, where it is found, let it be ever so improper, and likewise by proper instruments, such as are generally used by surgeons in their dissections, and not by coarse and rude knives and scissors, which may mangle and tear the body, but cannot ascertain the cause of its death.[19]

The Australian pathologist, Dr James Edward Neild, likewise emphasised in his lectures on forensic medicine that the manner in which a post-mortem examination was conducted was important for ensuring the efficacy of the death investigation. Neild went further than Farr in presenting the forensic gaze as central to how the examination ascertained the true cause of a death. To search '[i]nto those dark mysterious chambers of the soul where bad passions lurk, and evil desires spawn and multiply', Neild intimated, surgeons required the precision of a microscopical gaze.[20] The purpose of the science of medical jurisprudence was to bring to light what the *judicial* eye could not see, by applying rigorous scientific techniques of observation to reveal what was concealed by hidden, secret crimes.

In *The Birth of the Clinic*, Michel Foucault describes how a renewed interest in pathological anatomy in the nineteenth century reorganised the 'medical gaze'. The anatomical view of the corpse, which opened up the body to reveal its 'tissular surfaces', modified those techniques of observation that were intrinsic to the clinical experience. The epistemology of the gaze consisted of unfurling the body, unfolding its dark crevices, tracing signs of morbidity in organs and tissue, while constructing an ideal type of the non-diseased body. Foucault claims that Western medicine was founded upon an empiricism of the corpse,

previously been claimed by medico-legal historians: 'Medical Knowledge and the Early Modern English Coroner's Inquest' (2010) 23(3) *Social History of Medicine* 475. On the history of forensic medicine in England, see Catherine Crawford, 'Legalizing Medicine: Early Modern Legal Systems and the Growth of Medico-Legal Knowledge', in Michael Clark and Catherine Crawford (eds), *Legal Medicine in History* (Cambridge University Press, 1994).

19   Samuel Farr, 'Elements of Medical Jurisprudence', in Thomas Cooper (ed), *Tracts on Medical Jurisprudence* (James Webster, [1788] 1819 edn) 43. The first original treatise published in English was by George E. Male, *An Epitome of Juridical or Forensic Medicine; for the use of Medical Men, Coroners, and Barristers* (T. and G. Underwood, 1816).

20   James Edward Neild, 'Introductory Lecture to the Course of Forensic Medicine, 12 March 1866' (1866) 5 *Australian Medical Journal* 144, 153. In his *Address delivered to the annual meeting of the Victorian branch of The British Medical Association, 28 July 1882* (Stillwell & Co., 1882) 8, he wrote that 'the study of disease is most completely pursued in the dead-house, and that the progress of medicine is intimately associated with practical pathology'.

which constructed the body as measurable, cartographic and normative. 'Medical rationality', he writes,

> plunges into the marvelous density of perception, offering the grain of things as the first face of truth, with their colours, their spots, their hardness, their adherence. The breadth of the experiment seems to be identified with the domain of the careful gaze, and of an empirical vigilance receptive only to the evidence of visible contents. The eye becomes the depository and source of clarity; it has the power to bring a truth to light that it receives only to the extent that it has brought it to light; as it opens, the eye first opens the truth: a flexion that marks the transition from the world of classical clarity – from the 'enlightenment' – to the nineteenth century.[21]

Metaphors of light and dark infused the science of forensic medicine in the nineteenth century. The corpse was represented as a black box of obscurity, while the autopsy was depicted as a beacon of light. Central to this interplay of chiaroscuro was the forensic gaze. Only a meticulous 'empirical' gaze could peel away the shrouds of the body to uncover the true cause of a death, to unearth the bile of a crime. The clinical 'glance', particularly its narrow focus on a discrete area of the corpse, was no match for the attentive, precise, yet totalising forensic gaze. The technology of this gaze thus occupied a central role in the development of medical jurisprudence in England. The anatomical view of the corpse instituted a new mode of seeing. It projected the corpse as a dark entity which could only be enlightened through modern techniques of observation.

Super visum corporis (or 'upon the view of the body')

References to the legal custom of *super visum corporis* first appeared in Sir Matthew Hale's *History of the Pleas of the Crown* in a passage outlining the proceedings in the twelfth and thirteenth centuries of the law of *murdrorum*.[22] In the Middle Ages, *Lex Murdrorum* imposed amercements on townships for the slaying of Normans or the concealment of a Norman corpse within the vicinity of the town, regardless of whether death was caused by violence or accident.[23] It also required coroners to hold an inquest 'upon the view of the body', to confirm whether the corpse in question belonged to an English or Norman person. This was the case especially in cases where the town disputed the initial findings of the coroner. Coroners were required under this law to provide the community with

---

21 Michel Foucault, *The Birth of the Clinic: An Archaeology of Medical Perception* (Alan Sheridan trans, Routledge Classics, 2003) xiv.
22 Sir Matthew Hale, *Historia Placitorum Coronæ: The History of the Pleas of the Crown, Volume I* (E. and R. Nutt, and R. Gosling, 1736).
23 Jill McKeough, 'Origins of the Coronial Jurisdiction' (1983) 6 *University of New South Wales Law Journal* 191, 202.

an opportunity to prove the 'presentment of Englishry'. In other words, the legal presumption was that all unidentified corpses belonged to Normans and the onus of proving Englishry fell upon the townships that refused to be amerced. Hence, to view the body before a jury of peers was an important administrative procedure designed to prevent the community from burying the corpse prior to the arrival of the coroner and to prevent coroners from extorting fines for the discovery of a non-existent corpse.

While *Lex Murdrorum* was abolished in the fourteenth century, the custom of *super visum corporis* remained imperative until at least the early twentieth century. This was mainly due to the emphasis placed in *De Officio Coronatoris* (1276) on the coroner's duty to view all wounds when investigating a sudden, violent, or unnatural death. Yet, prior to the nineteenth century, the English inquest lacked any *formal* forensic component and provided little room for the presentation of medical evidence.[24] Although the coroner, as Jill McKeough opines, 'no doubt became quite proficient at recognising certain causes of death', during the Middle Ages, the performance of *super visum corporis* merely involved a 'casual glance' over the surface of the corpse.[25] And it was certainly irreducible to the thorough gaze demanded by the rigour of forensic medicine. *Super visum corporis*, as Hale wrote, was limited to an *external* examination of the dead body, an investigation of 'the place, length and depth of the wound'.[26]

The development of forensic medicine as a sub-discipline of human anatomy transformed the procedures for viewing the corpse. Prior to this shift, a

---

24 Ian Freckelton and David Ranson write that *An Act to Provide for the Attendance and Remuneration of Medical Witnesses at Coroners Inquests 1836* (UK), which enabled coroners to compel medical practitioners to conduct an autopsy and remunerate them for presenting evidence at an inquest, 'marked an important stage in the acceptance of forensic medicine and also constituted an acknowledgement of the significance of its contribution to scientifically conducted coroner's inquests': *Death Investigation and the Coroner's Inquest* (Oxford University Press, 2006) 16. However, it is noteworthy that the

> auditing of the coroner's affairs not only reduced the number of inquests, but effectively restricted the use, by the coroner, of expert witnesses which might have assisted him in his investigations. (If an expert witness was called by the coroner, the fee paid to the witness by the coroner might not be reimbursed if the justices deemed the evidence unnecessary, so that there was little incentive for coroners to be thorough and detailed in their investigation as such an approach might lead to considerable personal financial loss).
> David Ranson, 'The Role of the Pathologist', in Hugh Selby (ed), *The Aftermath of Death: Coronials* (Federation Press, 1992) 83

The *County Coroners Act 1860* (UK) overturned the practice of requiring coroners to self-fund the costs of conducting an inquest.

25 McKeough, above n 23, 203.

26 Sir Matthew Hale, *Historia Placitorum Coronæ: The History of the Pleas of the Crown, Volume II* (E. and R. Nutt, and R. Gosling, 1736) 58.

simple glance by the coroner (and jury) was sufficient to fulfil the duty to ascertain the manner of death. But with the advent of the science of medical jurisprudence, only an internal post-mortem examination of the body was deemed adequate to adduce the causes of death. By the late nineteenth century then, the forensic gaze had become integral to the way the coroner conducted a view of the dead body. It was no longer permissible for the coroner to superficially glance at marks of violence on the surface of the corpse. To be sure, the inquest could only lawfully proceed where the coroner could access the precision of the forensic gaze. In *R v Ferrand*, Justice Best reprimanded a coroner for simply looking at the face of the corpse, which did not provide him with 'a sufficient view of the body to give him authority to proceed'.[27] What was at stake in the transformation of the manner of viewing the corpse was precisely the scope of the jurisdiction of the coroner: 'a coroner has no manner of power to take an inquisition of death without a view of the body, and that any such inquest taken by him without a view is merely void'.[28]

Following the publication in 1814 of Mathieu Orfila's treatise on modern toxicology, the deceased's body, particularly where there was suspicion of poisoning, demanded a thorough anatomical investigation.[29] Ian Burney has written extensively about how the phenomenon of poisoning and the development of a science of toxicology catalysed an increase in the number of autopsies conducted as part of the English inquest in the latter half of the nineteenth century.[30] This was also the case in the Australian colonies, where post-mortem examinations were not regularly performed until the 1860s and 1870s. The first coroner of Melbourne, for example, rarely conducted autopsies on corpses under his jurisdiction. It was not that Dr Wilmot rejected the science of medical jurisprudence; indeed, he often called for medical witnesses to attend his inquests. Rather, he believed that the post-mortem did not radically affect the *quality* of the view of the dead body. He expressed such sentiment in an article written for the first issue of the *Australian Medical Journal*, in which he stated that the 'morbid condition of parts with which the post-mortem examination makes us acquainted, unfolds to us rather the effects of disease than the disease itself'.[31] Much to the coroner's chagrin, the public did not share 'his airy manner of dismissing

27 *R v Ferrand* (1819) 3 B & ALD 260, 264 (Best J).
28 Ibid, 263 (Holroyd J).
29 See, e.g., G.A. Paton, 'The Development of Forensic Medicine', in John Barry and R.J. Wright-Smith (eds), *The Proceedings of the Medico-Legal Society of Victoria 1939–1940–1941, Volume IV* (Brown, Prior, Anderson, 1941) 243.
30 See further, Ian A. Burney, 'A Poisoning of No Substance: The Trials of Medico-Legal Proof in Mid-Victorian England' (1999) 38(1) *Journal of British Studies* 59 and Ian Burney, *Poison, Detection and the Victorian Imagination* (Manchester University Press, 2012).
31 William B. Wilmot, 'On the Principles of Pathology' (1856) 1(1) *Australian Medical Journal* 1, 2.

an inquest',[32] and his verdicts of 'found dead' or 'visitation of god' were often characterised in newspapers as imprecise.[33]

It was not until the appointment of his successor that the autopsy became a mandatory component of the death investigation process in Victoria:

> Human life was not very highly esteemed in those early days, and the 'crowner's quests' were not of a very searching character. Frequently they were quite perfunctory, no post-mortem examination being made. Dr. Youl insisted on a post-mortem examination in every case, in order that the cause of death might be set beyond a doubt, and he also made a careful inquiry into all the surrounding circumstances.[34]

Coroner Youl's insistence that an autopsy be performed before every inquest formed part of a larger project to reform the death investigation process in the city of Melbourne.[35] His reforms were guided by the ideal that 'the actual cause of death might be placed beyond doubt',[36] but they were also intended to revalorise the efficacy of the coronial office, particularly as it had come under attack from politicians, physicians and journalists. Youl's demand that coroners carefully examine the surrounding circumstances of death and meaningfully gaze into the internal cavities of the body was central to his understanding of how coroners should take care of the dead. He maintained that his office had a duty

---

32 'The Pictorial: Men of the Day – The City Coroner', *The Australasian* (Melbourne), 25 April 1891, 812.

33 See 'The Lame, The Halt and The Blind', *The Argus* (Melbourne), 12 October 1855, 4:

> No effective means of identification! – 'found drowned' – 'Visitation of God,' &c., &c., constituted the sort of verdicts which satisfied the community of the freedom from all suspicion of foul play, and yet left the question 'open'; in case of any fresh light being thrown upon it. Vagueness and uncertainty were the grand characteristics of the decisions of those days.

34 *The Argus* (Melbourne), 7 August 1897, 10. Youl's belief in the necessity of post-mortem examinations was shared by the first coroner of New Zealand, Dr John Johnston. Johnston insisted on conducting post-mortem examinations where the death was not 'very apparent': A.J. Johnston, *A Handy Book for the Coroners of New Zealand* (Government Printer, 1868) 12.

35 Youl conducted over 12,000 inquests across his career: 'Obituary', *British Medical Journal* (London), 25 September 1897, 824. Andrew Brown-May and Simon Cooke explain that,

> [i]n a sample of inquests ... examined from the morgue in 1859, 42% included a *post-mortem* examination. In 1869–70 that figure had risen to 82%, and a medical examination of the body – without opening it – was held in a further 5% of cases.
>
> Brown-May and Cooke, above n 6, 15

36 'The Pictorial: Men of the Day: The City Coroner', *The Australasian* (Melbourne), 25 April 1891, 812.

to ensure that all lines of inquiry had been exhausted, and all possible theories had been excluded, when investigating the truth of what caused a death, which could only be determined by adopting the technology of the gaze.

In the third edition of a *Manual for Coroners and Magistrates in New South Wales*, which was published in 1895, Thomas MacNevin reiterated that the forensic gaze was a necessary condition of how the coroner conducted a view of the dead body. In the absence of a rigorous examination of the corpse, MacNevin declared, the inquest was void: 'If the body cannot be viewed, the Coroner can do nothing'.[37] Hence, the authority of the coroner could only materialise where the corpse could be observed. This view was influenced by Hale's earlier assertion in the seventeenth century that '[t]he coroner cannot take an inquisition but upon the view of the body, and if he doth, such inquisition is void; and the reason is, because oftentimes much of the evidence ariseth upon the view'.[38] However, where MacNevin differed from Hale was in delimiting precise procedures for charging the coronial jury both before and after they have observed the body, and instructing coroners in how the body should be viewed. In other words, the coroner's jurisdiction depended as much on viewing the corpse as on the *quality* of that view:

> The Coroner will then go with the jury and examine the body *minutely* and *carefully*. ... The position of the body, any external marks of violence, the state of the clothes, the position, appearance, and direction of the wounds (if any), and the relation of the surrounding objects, as well as any stains, marks, or appearances of blood, whether on the clothes of the deceased or near to the place where the body was first discovered, should be observed and accurately noted.[39]

The problem of the coroner and jury paying inadequate attention to the manner in which they viewed the body was bound to the question of the scope of coronial jurisdiction. The requirement that a corpse be viewed by the coroner prior to an inquest was not enough by the end of the nineteenth century to authorise proceeding with the hearing. The procedures of *super visum corporis* necessarily involved a *forensic* view, which was not only attuned to whether a post-mortem examination might be required – which inevitably was to be carried out in every case of sudden, violent, or unnatural death – but would also take in the scene of death prudently, holistically and judiciously. The manual unambiguously rejected the perfunctory glance that was conducted prior to the development of

---

37 Thomas E. MacNevin, *Manual for Coroners and Magistrates in New South Wales: Being a Practical Guide to the Proceedings of the Coroner's Court and to the Holding of Magisterial Inquiries in Lieu of Inquests by Justices of the Peace* (Charles Potter, Government Printer, 3rd edn, 1895) 24.

38 Hale, above n 26. This was confirmed in *R v Clerk* (1702) 2 Salk 277; 91 ER 328.

39 MacNevin, above n 37, 36 (emphasis added).

a science of forensic medicine. It instructed coroners to replace it with a rigorous, piercing, objective gaze. By the early twentieth century, coronial manuals stressed that 'a mere external manipulation of a body does *not* constitute a *post-mortem* examination'.[40] For the technology of the gaze had become an integral aspect of the death investigation process.

The problem of the coroner's view of the corpse

Modifications to procedures for viewing the corpse in the late nineteenth century catalysed jurisdictional conflicts between coroners, lawyers and physicians. While much has been written about the history of these disputes in England, little has been said about how they played out in Australia.[41] Local disputes between legal and medical professionals led to calls in the Australian colonies for the gradual attenuation of the view, the dissipation of the coronial jury and the development of forensic medicine as a distinct vocation with specialised training. In addition, politicians and journalists openly questioned the purpose, relevance and efficacy of the office of coroner. I suggest that these conflicts were produced by, but also reinforced, the reliance on the forensic gaze during the death investigation process. The question of who should assume responsibility for the care of the dead was discussed at length during these debates by the medical profession, who desired to retain the corpse as a technical-scientific object of medical knowledge, and by the legal profession, who saw in the history of the coronership the cultivation of relations that properly belonged to law.

While post-mortem examinations were rarely conducted on dead bodies in England prior to the nineteenth century, when they did occur medical practitioners ably assisted coroners. It was thus not surprising that, in Australia, autopsies were conducted by local physicians even though the majority of coroners were medically trained.[42] It was not until the 1860s that this procedure began to change, when legislation was enacted to prohibit physicians from conducting autopsies on their own patients, particularly when the coroner had grounds to suspect that the physician may have contributed to their patient's death.[43] This enraged medical professionals, who maintained that only the physician, who

---

40 Henry Storry Hawkins and Henry Giles Shaw, *Manual for Coroners and Magistrates in New South Wales* (W.A. Gullick, 1914) 39 (emphasis in original).

41 On the history of professional disputes in England, see J.D.J. Harvard, *The Detection of Secret Homicide: A Study of the Medico-legal System of Investigation of Sudden and Unexplained Deaths* (Macmillan, 1960); Ian A. Burney, *Bodies of Evidence: Medicine and the Politics of the English Inquest 1830–1926* (Johns Hopkins University Press, 2000); Joe Sim and Tony Ward, 'The Magistrate of the Poor? Coroners and Deaths in Custody in Nineteenth-Century England', in Clark and Crawford, above n 18.

42 This was not the case in England, where the majority of coroners were educated in law: see Ian A. Burney, 'Viewing Bodies: Medicine, Public Order, and English Inquest Practice' (1994) 2(1) *Configurations* 33.

43 The coroner was authorised under Section 16 of the *Medical Practitioners Statute 1865* (Vic) to

possessed intimate knowledge of their patient's medical history, could determine the true cause of death. They were also opposed to changes in such procedures, because the post-mortem was a source of extra income for medical practitioners in the Australian colonies. Early criticism of the office of coroner, then, was directed towards this legal prohibition, which many considered an affront to the doctor's ethical obligation to care for their patients.

Much to the chagrin of the medical profession, early coroners went further by lobbying governments for the creation of the role of medical jurist. In the 1860s, coroners called for governments to enact an office of the coroner's surgeon or government pathologist, as it later became known, who would be solely responsible for performing all post-mortems in a specific district. They lobbied for the creation of this role because they believed that it would mitigate conflicts of interest as well as improve the efficacy of the inquest.[44] However, for medical practitioners, this represented another attempt by law to usurp their jurisdiction to take care of the dead. In 1868, for example, a meeting of physicians complained that

> the appointment of a Government pathologist and medical jurist is a novel idea in the British empire, and would dissever the study of health from disease, supersede the vocation of the general medical practitioner, and divest him of his legitimate position and responsibility; lower him in the public estimation; and subject him to inquisitorial investigations, ruinously unnecessary inquests and harrowing to the feelings of the public.[45]

> direct any legally qualified medical practitioner to perform a *post-mortem* examination of the body of the deceased either with or without an analysis of the contents of the stomach or intestines. Provided that if in any case it appears to the coroner or justice that the death of such deceased person was probably caused partly or entirely by the improper or negligent treatment of any medical practitioner or other person then such practitioner or other person shall not be allowed to perform or assist at any such examination or analysis although he shall in every such case be allowed to be present thereat.

44 In correspondence between Coroner Wilmot and the Colonial Secretary, it was suggested that the role of medical jurist was required for the Colony of Victoria because physicians, while desiring to perform autopsies on their own patients, were often unable to do so 'within a proper time' and frequently left the corpse to decompose until a post-mortem was no longer valuable: Correspondence between Coroner Wilmot and Colonial Secretary, 9 July 1853, PROV, VPRS 1189, P0, UNIT 146, Item C53/6852.

45 Stephen Cordner and Fiona Leahy, 'Forensic Medicine and the Supreme Court', in Simon Smith (ed), *Judging for the People: A Social History of the Supreme Court in Victoria 1841–2016* (Allen & Unwin, 2016) 249. In 1862, 335 medical practitioners were registered in the Colony of Victoria, while, by 1881, this number had increased to 454. The tripartite division of the medical profession in England into colleges of physicians, surgeons and apothecaries was not replicated in the Australian colonies. Instead, all university-trained practitioners were listed on a single register. This difference is important to note when accounting for how medicine was institutionalised in the colonies. The Medical Society of Victoria was established

They were concerned, as Stephen Cordner and Fiona Leahy point out, that 'a government-appointed medical jurist, would operate as a "medical policeman", seeking to expose incompetent medical practitioners to public embarrassment and even the risk of criminal prosecution'.[46] The rise of medical specialisation in Australia was thus seen as threatening to expose the work of the physician to the vagaries of law and undermine the long-standing independence of the medical societies in adjudicating allegations of misconduct and malpractice.[47] While the exclusion of medical practitioners from the death investigation process was undoubtedly a source of discontent for many in the profession, by the same token they were dissatisfied with their inclusion within the scope of the coronial inquiry.[48]

The future of the coronership in Australia was discussed at length by politicians, lawyers, journalists and physicians in the 1870s and 1880s. Several politicians argued that the office should be abolished because 'medical men were the worst men to judge the cases which they had to deal with'.[49] They claimed that trained lawyers, such as police magistrates or justices of the peace, should be solely responsible for conducting death investigations.[50] Such debates referenced fiery discussions that took place much earlier in England about the suitability of

---

in 1855, while the British Medical Association opened an outpost in the colony in 1879: see further, Milton J. Lewis, 'Medicine in Colonial Australia, 1788–1900' (2014) 201(1) *Medical Journal of Australia* S5, S5–S6.

46 Ibid, 246. The authors explain that, in 1868, Coroner Youl raised the possibility of employing James Edward Neild as the first medical jurist for the Colony of Victoria at a meeting of the committee of the Melbourne Hospital. His view

> was supported by the hospital committee and the council of the university. Following formal endorsement by the Medical Society, it was further agreed to by the Minister of Justice. However, opponents of the scheme in the medical profession made a strong representation to the minister and ultimately defeated the proposal.
>
> Ibid, 248

47 For an insightful analysis of how this rivalry unfolded in Paris, see Jonathan Strauss, *Human Remains: Medicine, Death, and Desire in Nineteenth-Century Paris* (Fordham University Press, 2012) Ch 1.

48 While coronial investigations into deaths in hospitals and asylums were lauded in the media and led to numerous reforms in the provision of medical care in Australia, the coroner's 'vigorous attack upon the authorities for their neglect of sanitary precautions' was not equally well regarded by the medical professions: 'Death of Dr Youl', *The Argus* (Melbourne), 7 August 1897, 10.

49 Mr Bent quoted from 'Debate on Coroners in the Legislative Assembly Tuesday 2 October 1877' (1877) 22(10) *Australian Medical Journal* 305, 305.

50 This is despite the fact that, 20 years earlier, when criticised for only appointing medically trained coroners in the Colony of Victoria, the Attorney-General opined that 'legal knowledge was very little wanted in investigating causes of death … Medical men were the best of all qualified to act as coroners': Victoria, *Parliamentary Debates*, Legislative Assembly, 16 June 1857, 809 (Mr Owens). See also for a similar opinion in England, George Lowther, 'A Knowledge of Medicine an Essential Requirement in a Coroner' (1839) 1 *The Lancet* 578.

electing a physician to the role of coroner. The election of Dr Thomas Wakley to the coronership for Middlesex in 1839 was underpinned by a fierce rivalry between justices of the peace and members of the medical profession. Both viewed the other as lacking objectivity, inherently biased and conspicuously partisan. For justices of the peace, Wakley's election to the post threatened to 'introduce a "pernicious" form of professional prejudice', while medical practitioners protested that lawyers failed the public by conducting imprecise, unscientific inquests.[51] Nonetheless, these debates took a different turn in the Australian colonies, where coroners were medically trained and appointed rather than elected to their role.

Differences between the qualifications of coroners in Australia and England help explain why medical societies in the colonies adopted a somewhat contradictory attitude towards the continued existence of the coronial office. On the one hand, the medically trained coroner symbolised a betrayal of loyalty. The stench of law tainted what may have appeared to the medical profession to offer a potentially partisan role. Despite the possibility of medically trained office-holders positively influencing the conduct of death investigations, it was feared that a medical rather than legal coroner would further encroach on the jurisdiction of the physician. It was for such reasons that medical societies lobbied colonial governments to abolish the office of coroner and quash the proposal to appoint a medical jurist. But on the off chance that their efforts were not successful, the appointment of medically trained office-holders to the post was to be preferred over the selection of lawyers, as a means to prevent the coroner from obtrusively inquiring into the way physicians practised their craft. Sections of the medical profession thus lobbied in the alternative for the substitution of the coronial office with a system of 'medical examiners', which was popular at the time in continental Europe and some states of America.[52]

This two-pronged attack led to appeals in Parliament and elsewhere to split the duties of the coroner between justices of the peace and medical examiners. Numerous reasons were provided to justify this proposition, but what was missing in such calls – even in those from the medical profession, who opposed the continued existence of the coroner's office and any proposal to create an office for a medical jurist – was an acknowledgement that the conduct of inquests had substantially changed throughout the nineteenth century. The proliferation of

---

51 Burney, *Bodies of Evidence*, above n 41, 19. Thomas Wakley was a physician, founder and editor of *The Lancet* and a member of Parliament. Alongside William Farr, he was credited for the enactment of a civil death registration system in 1836: *An Act for Registering Births, Deaths and Marriages in England 1836* (UK).

52 See R.R. Scholl, 'Coroner's Inquests', in Barry and Wright-Smith, *The Proceedings of the Medico-Legal Society of Victoria 1939–1940-1941, Volume IV* (Brown, Prior, Anderson, 1941) 186. This argument was considered by a Departmental Committee in England in 1936 and rejected as 'not practicable or desirable in the present organization of the legal system'. However, the committee recommended limiting the coronership to the appointment of solicitors or barristers.

the forensic gaze was both a cause *and* an effect of jurisdictional conflict between lawyers and physicians about who should assume responsibility for caring for the dead. In other words, the criticism from medical societies that the inquest lacked scientific expertise, that it was inherently unscientific, was premised on an anachronistic interpretation of *super visum corporis*. By the time the future of the coronership was being debated among politicians, the press and in the community, the casual glance of the corpse had been utterly rejected by coroners. Professional manuals conveyed a coroner already equipped with the technology of the forensic gaze.

That the death investigation process was already undergoing transformation as calls were being made to abolish the office of coroner is further exemplified by examining the criticism levelled against the role of the coronial jury. In the late nineteenth century, the coronial jury was variously denounced by politicians as careless, slavish and useless.[53] They alleged that jurors were a wasteful expense, servile to the testimony of medical opinion and that they routinely deferred to the coroner to make findings on their behalf.[54] What was of most concern, though, was the jurors' view of the corpse, even where, as I discussed at the beginning of this chapter, it was mediated by a glass screen. In fact, the screen itself was thought to have rendered the duty of *super visum corporis* potentially useless.[55] To be sure, the purpose of a juror's view was questioned at length once post-mortems were regularly performed to diagnose the manner of death, and the view was depicted by many as morally, physically and spiritually unsettling.[56] It was seen as deleterious, misinformed and yet also too interested, which could in turn lead to perverse findings:

> On several occasions the view of the body by the jury, even when a Medical Practitioner has made a post-mortem examination, has been found to be misleading to the Inquiry. Post-mortem staining has been mistaken for

53 See Victoria, *Parliamentary Debates*, 'Coroners' Jurors', Legislative Assembly, 31 August 1887; Victoria, *Parliamentary Debates*, 'Coroners Juries Law Amendment Bill', Legislative Assembly, 11 October 1887; Victoria, *Parliamentary Debates*, 'Coroners Act Further Amendment Bill', Legislative Assembly, 1 April 1903.

54 This point was made by Burney in reference to England, but it also applied to the Australian colonies: see Burney, 'Viewing Bodies', above n 42, 39. For example, Mr Gaunson described the inquest as an expensive farce: 'jurors and witnesses were crammed up into small rooms, and the coroners might be seen with jugs of beer at their sides and large bowled pipes in their mouths': 'Debate on Coroners in the Legislative Assembly Tuesday 2 October 1877' (1877) 22(10) *Australian Medical Journal* 305, 306.

55 Ibid. The Secretary of the British Medical Defense Union claimed before a Department Committee in 1909 that '[y]ou may have had a dummy there so far as being able to assert anything as to the cause of death by that view is concerned. It is an absolute farce'.

56 See Scholl, above n 52, 195–196: 'The risk of the coroner being tricked into a fictitious inquest seems so remote, provided reasonable evidence is forthcoming of the existence of the body, that the 'view', occasioning the expenditure of time, and sometimes of money … should be dispersed with'.

bruises; the opening of the head has been mistaken for fractures of the skull, and it has been with the greatest difficulty that the matter has been explained to the Jury.[57]

The jury's view of the corpse was considered at best ignorant, and at worst misleading. 'Confusion, misreadings, and inaccurate verdicts', writes Burney, 'were the inevitable result of an uninitiated set of visual interpreters of the body, whose tendency to fix on meaningless external signs reflected an archaic order of knowledge'.[58] The coronial jury first became optional in parts of Australia in 1911, while in England it was no longer mandatory for all deaths from 1926.[59] Simon Cooke suggests that the demise of the coronial jury resulted in the loss of the inquest's civic purpose; it 'became increasingly professionalised and bureaucratised'.[60] I will return to the question of bureaucracy in the following chapter, but for now it suffices to say that one interpretation of this history is that the abolition of the jury restricted the coroner's view of the corpse. For, without the jury, as Burney has suggested, the custom of *super visum corporis* was emptied of its legal significance. It became a historical curiosity of the medieval coroner and a legal artefact destined for a museum of medical jurisprudence. Initially, after the jury was dismissed, the coroner viewed the body alone; while later in the twentieth century this task was left to the forensic pathologist.[61]

I have so far documented how modifications to procedures for viewing the corpse became a contentious point of disagreement between coroners, lawyers and physicians in the late nineteenth century. This issue was precariously positioned between the truth claims made by the medical expert, bolstered by a belief in the objectivity of the autopsy, and the interpretations of a lay jury that was often perceived to be subject to the whims of the coroner and reliant on

57 'Annual Reports of the Coroner's Society (1897–98) 48' quoted in Burney, 'Viewing Bodies', above n 42, 40. While the British Coroner's Society ultimately rejected proposals to restrict the view of the jury, Burney writes that it was still considered by others as 'an intrusive outrage, a sign of residual barbarity out of place in the modern world; it was a sanitarian's nightmare; finally, it was a source of profane interference with the efficient and purposeful production of scientific knowledge': Ibid, 36.

58 Ibid, 40.

59 Section 8 of *Coroners Act 1911* (Vic) allowed coroners to hold inquests without summoning a jury. See also Section 3 of the *Coroners (Amendment) Act 1926* (UK), which made the jury optional in certain circumstances. It remained compulsory, for example, where there was suspicion of murder or manslaughter.

60 Simon Cooke, *Inquests* (2008) The Encyclopedia of Melbourne Online. www.emelbourne.net.au/biogs/EM00756b.htm. See also Simon Cooke, *Secret Sorrows: A Social History of Suicide in Victoria, 1824–1921* (PhD Thesis, University of Melbourne, 1998) 77.

61 Victoria, *Parliamentary Debates*, 'Coroners' Law Consolidation and Amendment Bill', Legislative Council, 1 August 1911, 356.

the conjecture of circumstantial evidence.[62] In *Bodies of Evidence*, Burney characterises this as an epistemological crisis and argues that it conditioned the medicalisation of the inquest. He further asserts that the emergence of medical expertise challenged the long-standing notion of the inquest as a bulwark of public liberties and foreshadowed the demise of civic participation in the death investigation process.[63] Foucault offers a similar argument about the privileging of expert evidence as statements of truth, the medicalisation of judicial institutions and calls for the demise of the jury in nineteenth-century France.[64] While I do not deny that the development of forensic medicine 'constituted an attempt to recast inquests as primarily scientific events',[65] the positioning of the

---

62 '[S]cientific objectivity … emerged in the mid-nineteenth century and in a matter of decades became established not only as a scientific norm but also as a set of practices': Lorraine Daston and Peter Galison, *Objectivity* (Zone Books, 2010) 27. Belinda Carpenter and Gordon Tait contend that, during this period, the hermeneutics of law was often 'regarded as inferior mechanisms of truth-assessment': 'The Autopsy Imperative: Medicine, Law and the Coronial Investigation' (2010) 31 *Journal of Medical Humanities* 205, 210–211. That being said, medical science, in particular forensic medicine, was challenged by its own subjectivity, insofar as the autopsy was a contextual practice that involved the making of certain interested decisions over others. The subjectivity of medical knowledges did not, as Carpenter and Tait write, 'immediately translate into the necessary production of medical truth, and from there to a concomitant clear and infallible answer to all the questions raised in the coronial investigation'. Indeed, the laws of evidence did not empower the coroner to simply accept the truth claims made by medical experts. Rather, the coroner could allow them to provide an 'opinion' subject to cross-examination, even if their opinion did not originate from their direct observation of an event, but was mediated by their interactions with patients and other professionals. For more information on the history of epistemological conflicts between coronial law and forensic medicine in Australia, see William Ramsay Smith, *Medical Jurisprudence from the Judicial Standpoint* (Stevens and Sons, 1913); Crawford Henry Mollison, *Lectures on Forensic Medicine* (University of Melbourne, 1921).

63 He writes that the increased reliance on medical specialisation at inquests problematised the role of the coroner and the jury as important 'check[s]' on the growing influence of a centralizing administrative apparatus': Burney, *Bodies of Evidence*, above n 41, 5. For more on how the coronial office provided oversight on governmental abuse, see Ian A. Burney, 'Making Room at the Public Bar: Coroners Inquests, Medical Knowledge and the Politics of the Constitution in Early-Nineteenth-Century England', in James Vernon (ed), *Re-reading the Constitution: New Narratives in the Political History of England's Long Nineteenth Century* (Cambridge University Press, 1996).

64 Michel Foucault, *Abnormal: Lectures at the Collège de France 1974–1975* (Graham Burchell trans, Picador, 2003) 11 and 39. Foucault explains that, towards the end of the nineteenth century,

> we hear quite serious proposals for the suppression of the jury. The jury, it is argued, [is made up of] people who are neither doctors nor judges and who consequently are competent neither in law nor in medicine. A jury of this kind can only be an obstacle, an opaque element, a resistant block within the judicial institution as it ought to be ideally. How would the true judicial institution be composed? It would be made up of a jury of experts under the juridical responsibility of a magistrate.

65 Burney, 'Viewing Bodies', above n 42, 33.

forensic gaze as intrinsic to the legal duty to view the corpse also affected how coroners assumed responsibility for taking care of the dead. To put this differently, the medicalisation of the coronial inquest was only one consequence of the binding of the gaze to the institutional life of coronial law. Another consequence was the revalorisation of the role of the coroner in maintaining legal relations with the dead.

The procedures for viewing the dead – whether they died suddenly, violently, or unnaturally – before or during an inquest was central to calls for the abolition of the coronership, the dissipation of the jury and its replacement with a system of medical examiners. I have shown how transformations of this view were not simply a result of, but were inherently associated with, jurisdictional conflicts between coroners, lawyers and physicians about who should assume responsibility for the care of the dead. Burney has interpreted these conflicts in England as evidence of the medicalisation of the coronial inquest. He argues that a consequence of this was the evacuation of the body, a process of 'decorporealization', which transformed the inquest into 'an expert-based, efficiency-oriented system of death management'.[66] The body effectively became an 'abstract' sign in a witness's deposition or pathologist's report, while the coronial inquest was reduced to 'a paper inquiry'.[67]

At the same time, Burney notes that the body could never completely vanish from the inquest, because its presence granted the coroner the authority to inquire into the who, when, where and how of a death. Indeed, *super visum corporis* had to be performed before the coroner could lawfully proceed with an inquisition. Even if they desired to remove themselves from the 'unsettling materiality' of death – Burney cites an example of an English coroner attempting to distinguish in a memoir the duties of office from the indignity of conducting an autopsy – their jurisdiction depended on conducting a view of the corpse.[68] What is more, as noted above, the jurisdiction of the coroner was contingent on the quality of that view, which is to say that it was reliant on the attachment of the forensic gaze to the conduct of coronial law.

It is to this extent that the transformations of *super visum corporis* in the Australian colonies did not simply result in the medicalisation of the coronial inquest. While the development of forensic medicine, the rise of the medical expert and the attenuation of the jury's view all problematised the jurisdiction of the coroner, it does not simply follow that the inquest was reduced to a 'paper inquiry', which conjures an image of paper-shuffling. On the contrary, Australian coroners were keen to quell any doubt that their reliance on expert evidence and the waning of the jury's view undermined their authority to

---

66 Ibid, 41.
67 Ibid, 42. See also '"The View" at Inquests', above n 2, 996: 'To omit the view by the coroner would be to throw away the one outward and visible sign by which his direct control is manifested'.
68 Ibid.

perform the duties of their office. What is most important then is not the question of whether the office of coroner lost exclusive control of the corpse – a question that is premised on the erroneous assumption that the coroner once possessed exclusive control – but rather how the technology of the gaze transformed the means through which coroners assumed responsibility for taking care of the dead.

To speak on behalf of the dead

Foucault demonstrates in *The Birth of the Clinic* how the transformation of the 'medical gaze', which was conditioned by the anatomist's interest in opening up corpses, and thereby proving that vitalism was the opposite of death, shaped '[a] new alliance ... between words and things, enabling one *to see* and *to say*'.[69] The gaze uncovered a truth of the body by expressing in language 'what for centuries had remained below the threshold of the visible and the expressible'.[70] It translated a vision of flesh and bones, tissue and organs, lesions and wounds into a language that could be spoken. The anatomical view of the corpse thus constituted both '[a] hearing gaze and a speaking gaze: clinical experience represents a moment of balance between speech and spectacle'.[71] The historical account outlined above of the transformations of *super visum corporis* in the late nineteenth century has similarly unravelled a new relationship between a visual technology and a rhetorical style. *Super visum corporis* was never simply a mode of seeing, it was much more than this, even in the Middle Ages. It was a technique of speaking, a device for translating anatomical wounds into an institutional narrative.

In the final section of this chapter, I examine how the forensic gaze emerged in the late nineteenth century as an oratorical device particular to the persona of the coroner. It became a rhetorical style through which the coroner assumed responsibility for speaking on behalf of the dead. The medical expert undoubtedly assumed greater authority at this point for viewing the corpse, most often alone in the mortuary. However, this does not mean that coroners no longer maintained legal relations with the dead. The attachment of the forensic gaze to the institutional life of coronial law instead transformed how coroners cultivated those relations. To speak on behalf of the dead became part of a repertoire of jurisdictional technologies through which the coroner assumed responsibility for taking care of the dead.

The idea that *super visum corporis* constituted a rhetorical style requires some explanation through a historical analysis of the theory of persona. In Latin, *persona* denotes a role, a character, a mask. In a wide-ranging essay on the history of the concept, Marcel Mauss notes that the link between persona and mask was

69 Foucault, above n 21, xiii (emphasis in original).
70 Ibid
71 Ibid, 142.

instituted by the Romans, even though many other ancient civilisations also developed an economy of masks.[72] Roman citizens adorned different masks in civil life, including ancestral, ritual, or dramatic masks, which corresponded to the occupation of different roles in society. But, more than this, masking attached a persona to a legal order. It did not so much reveal a 'true identity', a 'natural' human being behind the persona, as Edward Mussawir writes, but rather depicted 'a fragmented or non-totalized identity linked to a discrete civic function'.[73] Roman citizens performed the mask of their ancestors in order to appear before the law; they had to become the mask itself to present themselves in person, and to do so they had to fashion a specific persona that corresponded with the particularity of that juridical operation. 'Persona was not what one *is*, but what one *has*, like a faculty that, precisely for this reason, you could also lose'.[74]

The technology of the mask bound a persona to a juridical capacity in Ancient Rome. It was, as Connal Parsley explains, 'a device, *dispositif* or apparatus – through which a juridical relation to life comes to be engendered'.[75] The Latin

---

72 Marcel Mauss, 'A Category of the Human Mind: The Notion of Person; The Notion of Self' (W.D. Halls trans) in Michael Carrithers, Steven Collins and Steven Lukes (eds), *The Category of the Person: Anthropology, Philosophy, History* (Cambridge University Press, 1985) 15. For further analysis on the Greek origins of the mask, see Jean-Pierre Vernant and Françoise Frontisi-Ducroux, 'Features of the Mask in Ancient Greece', in Jean-Pierre Vernant and Pierre Vidal-Naquet, *Myth and Tragedy in Ancient Greece* (Janet Lloyd trans, Zone Books, 1988).

73 Edward Mussawir, *Jurisdiction in Deleuze: The Expression and Representation of Law* (Routledge, 2011) 31. In analytical jurisprudence, 'personality' is depicted as a legal fiction, an abstract concept or a technical artifice, yet one that is added to a 'natural' individual that pre-exists law: see Alexander Nekam, *The Personality Conception of the Legal Entity* (Harvard University Press, 1938); Martin Wolff, 'On the Nature of Legal Persons' (1938) 54 *The Law Quarterly Review* 494; Frederick Henry Lawson, 'The Creative Use of Legal Concepts' (1957) 32 *New York University Law Review* 907; Hans Kelsen, *Pure Theory of Law* (University of California Press, 1967). The person has been conceived of as a legal device, whereby a corporation, non-human animal or human being is endowed with 'a formal capacity to bear a right or duty': Ngaire Naffine, 'Who Are Law's Persons? From Cheshire Cats to Responsible Subjects' (2003) 66 *The Modern Law Review* 346, 352. For a critique of analytical conceptualisations of the legal person, see Peter Goodrich, 'Specula Laws: Image, Aesthetic and Common Law' (1991) 2(2) *Law and Critique* 233; Richard Mohr, 'Flesh and the Person' (2008) 29 *Australian Feminist Law Journal* 31; Roberto Esposito, 'The *Dispositif* of the Person' (2012) 8(1) *Law, Culture and The Humanities* 17; Edward Mussawir and Connal Parsley, 'The Law of Persons Today: At the Margins of Jurisprudence' (2017) 11(1) *Law and Humanities* 44.

74 Roberto Esposito, *Persons and Things: From the Body's Point of View* (Zakiya Hanafi trans, Polity Press, 2015) 30 (emphasis in original). Hence, Giorgio Agamben points out that '[t]he slave, inasmuch as he or she had neither ancestors, nor a mask, nor a name, likewise could not have a "persona," that is, a juridical capacity (*servus non habet personam*)': *Nudities* (David Kishik and Stefan Dedatella trans, Stanford University Press, 2011) 46 (emphasis in original).

75 Connal Parsley, 'The Mask and Agamben: The Transitional Juridical Techniques of Legal Relation' (2010) 14 *Law Text Culture* 12, 12.

persona was originally translated by Cicero from the Greek *prosopon* (the *imago*, or death mask), which signified both a face and a disguise. Hence, the mask attached a juridical capacity to the artifice of a face. The face is the mask that lies beyond the mask, when 'every mask [is] torn away ... there is retained the sense of the artificial'.[76] In *Leviathan*, Thomas Hobbes highlighted the ineluctability of the face, artifice and persona in the mask:

> The word Person is latine: insteed whereof the Greeks have [*prosopon*], which signifies the *Face*, as *Persona* in latine signifies the *disguise*, or *outward appearance* of a man, counterfeited on the Stage; and sometimes more particularly that part of it, which disguiseth the face, as a Mask or Visard: And from the Stage, hath been translated to any Representer of speech and action, as well in Tribunalls, as Theaters. So that a *Person*, is the same that an *Actor* is, both on the stage and in common Conversation; and to *Personate*, is to *Act*, or *Represent* himselfe, or an other; and he that acteth another is said to beare his Person, or act in his name.[77]

What is most interesting about this passage is how rhetoric occupied a central role for Romans in framing the discursive relationship between a face and a disguise.[78] This means that the mask was not simply an optical apparatus for appearing before the law *in person*, but a rhetorical device for *personification*, which, as Mauss further points out, derives from *per* and *sonare* – 'the mask through which (*per*) resounds the voice (of the actor)'.[79] The mask of persona was thus an apparatus of vocalisation, which translated the speech and actions of the actor from the theatrical stage to the legal forum. It was a means of translation – recalling Foucault's characterisation of the medical gaze in *The Birth of the Clinic* – for describing what one sees and hears in a legal idiom. In addition, the mask of persona was not only used to personify the actions or speech of oneself – that is to say, the voice of the actor before the Roman forum – but was also deployed as a figure of speech to personate an other. In this act of

---

76 Mauss, above n 72, 18.
77 Thomas Hobbes, 'Chapter XVI: Of Persons, Authors, and Things Personated', in C.B. Macpherson (ed), *Leviathan* (Penguin Books, 1968) 217 (emphases in original).
78 For a discussion of the rhetorical origins of legal discourse, see Peter Goodrich, 'Rhetoric as Jurisprudence: An Introduction to the Politics of Legal Language' (1984) 4 *Oxford Journal of Legal Studies* 84; Peter Goodrich, *Reading the Law: A Critical Introduction to Legal Method and Techniques* (Basil Blackwell, 1986); Michael H. Frost, *Introduction to Classical Legal Rhetoric: A Lost Heritage* (Routledge, 2005) and Victoria Wohl, *Law's Cosmos: Juridical Discourse in Athenian Forensic Oratory* (Cambridge University Press, 2010). It should be noted that forensic oratory was a distinctive genre of legal rhetoric in antiquity. It defined precepts for the formulation and presentation of arguments before a court of law.
79 Mauss, above n 72, 15. However, Mauss notes that this interpretation of *persona* emerged later than the mask, such that the Romans distinguished 'between *persona* and *persona muta*, the silent role in drama and mime'.

personification, the 'actor' was said to bear or adorn the other's persona through the performance of the mask.

Hobbes's description of the rhetorical functions of the mask of persona recalls the rhetorical device of *prosopopoeia*, which derives from the Greek word for *prosopon*. *Prosopopoeia* is a trope or figure of speech that enables a writer or orator to speak on behalf of another. Classic examples of the use of this trope can be found in Greek and Roman mythology where inanimate objects are bestowed with a human voice. In the *Institutes of the Orator*, Quintilian described *prosopopoeia* as the most difficult kind of rhetorical speech. It provided the Roman orator with techniques for speaking on behalf not only of others, but also of things. In fact, Quintilian specified the dead, who were incapable of appearing before the forum, as important beneficiaries of this device. The rhetorical figure enabled the orator to raise the dead from their graves and lend them the means of locution.[80] Integral to the use of this device was the mask of persona. The Roman orator could only speak for the dead before the law through the expression of a persona, through the performance of the mask, and, of course, with the rhetoric of *prosopopoeia*.

This brief history of persona as a masked performance is useful for examining how the transformations of *super visum corporis* affected the conduct of coroners in the late nineteenth century. I previously discussed the idea, which is usually offered by medical historians, that, towards the end of the nineteenth century, the coronial inquest had become truly medicalised and the forensic expert had assumed almost exclusive authority for the corpse. In contrast to this view, I propose that, far from diminishing the jurisdiction of the coroner, the proliferation of the forensic gaze transformed the way coroners maintained legal relations with the dead. If *super visum corporis* is conceptualised not only as a visual technology, but also as a rhetorical style particular to the persona of the coroner, I argue that it functioned as a device for the personification of the dead. In other words, what I first considered as a jurisdictional technology that historically belonged to the office of coroner, but then explored as an optical apparatus specific to the expression of a coronial persona, was never simply a mode of viewing the dead. Even when medical experts and others lampooned the coroner's (and the jury's) use of *super visum corporis* (or their duty to view the corpse), they fundamentally misinterpreted its rhetorical resonances, which for years had reverberated throughout the institutional life of coronial law. It is thus

---

80 Quintilian, *Institutes of the Orator, Volume 2* (J. Patsall trans, B. Law and J. Wilkie, 1774) Book 9, Chapter 2. On the orality of things, see Joel Snyder, '*Res Ipsa Loquitur*', in Lorraine Daston (ed), *Things That Talk: Object Lessons from Art and Science* (Zone Books, 2008); Patrick R. Crowley, 'Roman Death Masks and the Metaphorics of the Negative' (2016) 64 *Grey Room* 64. While the voices of the dead have not been subject to much academic analysis, the rhetoric of *prosopopoeia* has featured in critical and literary theory. See, e.g., Michael Riffaterre, 'Prosopopeia' (1985) 69 *Yale French Studies* 107; James J. Paxson, *The Poetics of Personification* (Cambridge University Press, 1994); Diana Fuss, 'Corpse Poem' (2003) 30(1) *Critical Inquiry* 1.

possible to conceive of the view as not simply a visual technology, but, through the rhetoric of *prosopopoeia*, an oratorical device for speaking on behalf of the dead.

The theatricality of the inquest should not come as a surprise to modern readers. The idea that coroners were actors among others in the unfurling tragedy of the inquest is not out of place with contemporary scholarship that depicts the dramatic functions of law. Indeed, Peter Goodrich has pointed out that the Greek root of persona (*prosopon*), the actor's mask, denotes a dual meaning of the person – through its inscription in the domains of law and theatre, and its appearance before a legal forum and public audience.[81] The notion that coroners speak for those who have no means of locution is also unsurprising given that such acts have now been imagined as an important aspect of the coronial jurisdiction.[82] What this chapter has offered, then, is a historical account of how coroners became capable of speaking for the dead; that is, what rhetorical devices enabled them to translate a view of the corpse, its wounds, lesions, spots and colours, into figures of speech. Speaking on behalf of the dead was one aspect of how coroners maintained legal relations with the dead, particularly when the development of forensic medicine – specifically the rise of the medical expert and forensic pathologist – threatened to usurp their jurisdiction to take care of the dead, which originated as far back as the twelfth century. I argue that this means that the medicalisation of the inquest is now considered to be only part of the story of how the forensic gaze transformed the role of the coroner in the death investigation process. The other part of the story is that the gaze was never simply an optical apparatus – it was an institutional practice, a legal technology and a rhetorical device that authorised the coroner to speak on behalf of the dead.

This chapter has examined how the transformation of *super visum corporis* in the late nineteenth century affected the way coroners maintained legal relations with the dead. It has suggested that the coroner's view of the corpse was never simply a mode of seeing, but also a rhetorical device for speaking on behalf of the dead. It has further contended that such oratorical techniques became integral to how coroners assumed responsibility for taking care of the dead. When Robert Pogue Harrison writes that '[w]e speak with the words of the dead', he means that 'the dead speak in and through the voices of the living. We inherit their words so as to lend them voice'.[83] *Super visum corporis* was a jurisdictional technology through which the coroner could lawfully speak with the words of

---

81 Peter Goodrich, 'The Theatre of Emblems: On the Optical Apparatus and the Investiture of Persons' (2012) 8(1) *Law, Culture and the Humanities* 47, 56. See also Marett Leiboff, 'Law, Muteness and the Theatrical in Law's Theatrical Presence' (2010) 14 *Law Text Culture* 384.

82 See, e.g., Stanley Myles MacKenzie Leslie, *Speaking for the Dead: Coroners, Institutional Structures and Risk Management* (PhD Thesis, Centre for Criminology and Sociolegal Studies, University of Toronto, 2011).

83 Robert Pogue Harrison, *The Dominion of the Dead* (The University of Chicago Press, 2003) 151.

the dead. This perhaps explains why '[w]ounds do not speak until and unless they have the voice and utterance of another'.[84] And this chapter has shown that they do not speak until they are attached to an institutional idiom of law. The technology of the gaze provided coroners with this idiom, and the rhetorical tools for translating a view of the corpse into an institutional narrative of death causation.

84 Nina Philadelphoff-Puren and Peter Rush, 'Fatal (F)laws: Law, Literature and Writing' (2003) 14 *Law and Critique* 191, 197.

3 The bureaucratic logic of office

In the preface to the third edition of the *Manual for Coroners and Magistrates in New South Wales*, Thomas E. MacNevin, the chief clerk of the Department of Justice, painstakingly enumerated all that had been altered from previous editions. The second edition, which was published in 1884, was shorter in length, contained fewer forms and excluded circulars. The third edition, which was published 11 years later, by contrast included all circulars issued to coroners in the colony since 1868. The reason for this was the necessity of addressing a 'failure on the part of Coroners in many instances' to adhere to procedure in completing forms and 'avoid, as far as possible, needless correspondence on the subject'.[1] Irregularities included coroners failing to affix jurats, attach seals, or properly compile forms, which, as MacNevin explained, 'materially affect[ed] the validity of the Inquest'.[2] The importance of ameliorating these procedural errors was explicitly noted as an objective of the manual:

> The first edition of this little work, which has been for some time out of print, was published in 1876, and distributed to the Coroners throughout the Colony for their guidance and assistance in the performance of the duties pertaining to their important office; and was rendered necessary in consequence of certain irregularities which occurred from time to time in connection with the holding of inquests in the county districts, and the defective manner in which the evidence of witnesses was taken, without, in

---

1 Thomas E. MacNevin, *Manual for Coroners and Magistrates in New South Wales: Being a Practical Guide to the Proceedings of the Coroner's Court and to the Holding of Magisterial Inquiries in Lieu of Inquests by Justices of the Peace* (Charles Potter, Government Printer, 3rd edn, 1895) ii. R.H. Matthews, who was a coroner and justice of the peace for New South Wales, published the first edition of the *Handbook to Magisterial Inquiries in New South Wales: Being a Practical Guide for Justices of the Peace in Holding Inquiries Respecting Deaths* in 1888. This handbook was similar in many respects to MacNevin's manual, which was appropriate considering that '[i]n the absence of a coroner, the police magistrate was authorized to conduct magisterial inquiries into sudden deaths, and if he was unavailable a local justice of the peace could act': Government Gazette (Sydney), No 22, 15 March, 1845, 300.
2 Ibid.

many instances, the necessary jurats being attached to each disposition, tending to affect the validity and materially to lessen the usefulness of these investigations.[3]

The manual was indispensable for coroners in the late nineteenth century. Alongside a bundle of forms, recognisances, certificates and warrants; a Bible; writing material and 'a sufficient quantity of foolscap paper',[4] coroners were encouraged to carry a copy of the manual to the site of death. In the absence of any formal training for the role – individuals seeking office were presumed to be 'acquainted in a general way with those duties'[5] – it guided coroners in the performance of their duties. While MacNevin recommended that all coroners should also familiarise themselves with 'Sir John Jervis on the Office and Duties of Coroners', he acknowledged that there was no equivalent textbook for the Australian colonies, and his manual, which was provided to all coroners upon their first appointment to the role, was the only substantive attempt to address the paucity of local knowledge.[6]

The coronial manual first appeared in England in the eighteenth century.[7] *The Coroner's Guide* (1756) and *Lex Coronatoria* (1761) were followed throughout the nineteenth century by several handbooks written specially for coroners, including John Impey's *The Office and Duty of Coroners* (1800), Sir John Jervis's *A Practical Treatise on the Office and Duties of Coroners* (1829), Richard Sewell's *A Treatise on the Law of Coroner* (1843) and William Baker's *A Practical Compendium* (1851).[8] While English handbooks initially served as

---

3 Thomas E. MacNevin, *Manual for Coroners and Magistrates in New South Wales: Being a Practical Guide to the Proceedings of the Coroner's Court and to the Holding of Magisterial Inquiries in Lieu of Inquests by Justices of the Peace* (Government Printer, 2nd edn, 1884) v.

4 MacNevin, *Manual for Coroners and Magistrates in New South Wales*, 3rd edn, above n 1, 24.

5 Ibid, 2.

6 Ibid.

7 The earliest handbook written for coroners was published in England in 1756. The first edition of this orphan work remains unknown. Multipurpose handbooks written for sheriffs, bailiffs, justices of the peace, constables and coroners were circulated prior to the eighteenth century: see Sir Anthony Fitzherbert, *In This Boke Is Conteyned ye Office of Shyryffes, Baylyffes of Lybertyes, Escheatours, Constables, & Coroners: And Sheweth What Euerye One of Them May Do by Vertue of Theyr Offyces Drawen Out of Bokes of the Comen Lawe & of the Statutes* (Wyllyam Powell, 1549); Richard J. Burn, *Justice of the Peace and Parish Officer, Volume II* (Sweet, Maxwell and Son, and Steven's and Norton, [1755] 1845 edn); John H. Plunkett, *The Australian Magistrate or A Guide to the Duties of a Justice of the Peace for the Colony of New South Wales* (Gazette Office, 1835).

8 *The Coroner's Guide: or, the Office and Duty of a Coroner: Containing Variety of Precedents, and Proper Instructions for Executing the Said Office. Compiled from the Best Authorities* (John Worrall, 1756); Edward Umfreville, *Lex Coronatoria: or, the Office and Duty of Coroners. In Three Parts. Wherein the Theory of the Office Is Distinctly Laid Down; and the Practice Illustrated* (R. Griffiths and T. Becket, 1761); John Impey, *The Office and Duty of Coroners* (J. Butterworth, 1800); Sir John Jervis, *A Practical Treatise on the Office and Duties of Coroners* (S. Sweet, 1829); Richard Clarke Sewell, *A Treatise on the Law of Coroner* (O. Richards, 1843); William Baker, *A Practical Compendium of the Recent Statutes, Cases and Decisions Affecting the Office of the Coroner* (Butterworths, 1851).

guidance for coroners in British colonies, over the course of the nineteenth century, specialised manuals written by local officers were disseminated to accommodate the particular circumstances of colonial life. In Canada, William F.A. Boys wrote *A Practical Treatise on the Office and Duties of Coroners in Upper Canada* (1864), while, in New Zealand, Alexander J. Johnston published *A Handy Book for the Coroners of New Zealand* (1868).[9] In Australia, P.S. Tomlins (1837) and Thomas E. MacNevin (1875) produced manuals for the colonies of Van Diemen's Land and New South Wales, respectively.[10] Manuals specifically written for coroners in British colonies differed from their English counterparts insofar as they addressed the problem of transmitting legal opinions and advice across vast colonial frontiers.

This chapter investigates how the coronial manual functioned as a technology of office in the late nineteenth and early twentieth centuries. The manual guided coroners in interpreting the scope of their jurisdiction, the performance of their duties and the proper administration of inquests. It technocratised the practices of coronial law and procedure, while offering guidance on how to fulfil the obligations of the coroner's office. The professional handbook was undoubtedly preoccupied with questions of technical knowledge, skill and expertise. This, however, does not mean that it was bereft of an ethics of responsibility. It held on to the question of responsibility by framing the death investigation process within a bureaucratic logic of office. This chapter shows how the coronial manual assumed an indispensable role in the formation of an ethical mindset towards the dead.

The technocratic manual

MacNevin's 'practical guide to the ordinary duties of their office'[11] was most useful for coroners located in the outlying districts of the Colony of New South Wales. The difficulty of obtaining access to circulars issued by the Department of Justice, opinions from the Crown Law Offices or forms distributed by the Government Printer contributed to the demand for a coronial manual. This

---

9 William F.A. Boys, *A Practical Treatise on the Office and Duties of Coroners in Upper Canada: With an Appendix of Forms* (W.C. Chewett & Co, 1864); Alexander James Johnston, *A Handy Book for the Coroners of New Zealand* (Government Printer, 1868). See also John G. Lee MD, *Hand-book for Coroners* (W. Brotherhead, 1881), which was published in Philadelphia, and set out coronial law and procedure in 38 states of America.

10 P.S. Tomlins, *The Coroner's Guide: A Summary of the Duties, Powers and Liabilities of Coroners from the Most Approved Authorities* (Van Diemen's Land, 1837); Thomas E. MacNevin, *Manual for Coroners and Magistrates in New South Wales: Being a Practical Guide to the Proceedings of the Coroner's Court and to the Holding of Magisterial Inquiries in Lieu of Inquests by Justices of the Peace* (Government Printer, 1st edn, 1875). See also *Acts Relating to Coroners and Instructions for the Guidance of Coroners on Justices Acting as Such* (Government Printer, 1890), which can be accessed from the Victoria Police Museum and Historical Services Unit; James Drysdale Brown, *A Short Manual: For the Guidance of Coroners, Deputy Coroners, and Justices Acting as Coroners, in Victoria* (Government Printer, 1911).

11 MacNevin, *Manual for Coroners and Magistrates in New South Wales*, 2nd edn, above n 3, v.

problem also influenced editorial decisions about what should be included and excluded in each edition. The second edition, for example, expanded the 'practical' component of the manual by adding opinions of the Crown Law Offices, revising the forms necessary for practice and, in general, redesigning the typographical arrangement of the book for the sake of professional convenience, but also to ensure consistency among the hotchpotch of coroners operating in different corners of the colony:

> It is hoped, in conclusion, that, in the absence of any regular text-book treating of the practice of the Coroner's Court in this Colony, this manual will be found to be a trustworthy guide to the ordinary duties of the Coroner, and that the simple arrangement of the duties incident to the office, attempted in the present compilation may in some measure facilitate the performance of those duties, secure uniformity of practice, and assist the Coroner, conveniently and effectually, to discharge the important functions of his office. If these desirable objects be attained, the time, labour, and study that have been devoted by the compiler in the endeavour to produce a practical handbook upon a very difficult subject will have been profitably applied.[12]

The third edition brought additional changes to the content and structure of the manual. The chief clerk rearranged the order of chapters, revised the substance of precedents and added a number of forms in an enlarged Appendix. Under 'Forms of Verdicts', he included death 'during operation under chloroform' (Form 19), 'from effects of injuries' (Forms 20 and 21), 'accidentally overlain by mother' (Form 22), 'from asphyxia at birth' (Form 23) and 'from poison self-administered' (Form 24). The number of forms of verdicts expanded from 25 to 32, which did not denote simply that individuals had invented new modes of dying over the course of a decade, but rather that content in the manual had been augmented by technical knowledge. The *technocratisation* of the manual is evident when comparing these forms to the categories of verdicts in treatises written by English coroners in the eighteenth century. For example, in *The Coroner's Guide*, which was first published in 1756, the anonymous author confined possible verdicts to felony (suicide or murder), mischance (act of god or act of man) and famine (poverty or pestilence).

Technical knowledge was not supplementary to judicial statements interpreting the scope of the coronial jurisdiction. In the late nineteenth century, the jurisdiction of the coroner was shaped by a combination of case law and legislation.[13] However, in the third edition of MacNevin's manual, circulars,

---

12 Ibid, vi. There were approximately 130 coroners spread throughout the colony in 1894: Hilary Golder, *High and Responsible Office: A History of the NSW Magistracy* (Oxford University Press, 1991) 118.

13 British coronial law was transported to the colony in 1788, although it was not until 1828 that particular duties of the coroner were codified. The relevant Act at the time of publication of the third edition of MacNevin's manual was *An Act for Adopting Certain Acts of Parliament passed during the Seventh and Eighth Years of His Present Majesty King George the*

forms and precedent were interspersed with case law and legislation throughout each chapter.[14] The procedures for conducting an inquest – from the proclamations made at the opening of court, to pro forma declarations for the swearing of witnesses, to the form of vouchers for the advanced payment of jurors – were not ancillary to legal judgments on the scope of the coroner's authority. The procedural aspects of form-filling were part of the official instructions that guided coroners in navigating their inheritance of coronial law. In other words, substantive law and administrative procedure were not represented in the manual as mutually exclusive aspects of the conduct of office. Rather the manual embedded techniques on how to interpret the scope of coronial jurisdiction, how to perform the ordinary duties of office and how to conduct inquest proceedings among judicial consideration of Australian and English case law and legislation.

Circulars, forms and precedent formed devices for the transmission and representation of the authority of the coroner. The circular, in particular, which transmitted the opinions, minutes and advice of the Attorney General or Minister of Justice, constituted such a device insofar as it was created with the aim of modifying procedural relations between the coroner and the Crown. For example, the Colonial Secretary's Office admonished coroners in 1845 for overstepping the limits of their territorial jurisdiction. It transmitted notice of its decision by the use of a circular that stated 'in future no Coroner is to act except within the Police district in which he may reside and for which he is understood as holding his appointment'.[15] Yet MacNevin noted in the Preface to the third edition of his manual that, by 1892, the Department of Justice had issued another circular expanding the territorial jurisdiction of the office to enable coroners, in special circumstances, to hold inquests beyond the limits of their appointed district. What is most interesting about MacNevin's decision to not simply quote from the circular, but transcribe both documents in the Appendix of the manual, is that it represented the document as a technology for delimiting the territorial jurisdiction of the coroner. Coronial jurisdiction did not pre-exist the issuing of official opinions on the matter; it materialised through the transcription and dissemination of circulars in the manual.

Circulars issued to coroners ranged from instructions for the mode of transmitting inquisition documents to processes for stamping official correspondence. The art of anthologising such devices in the Appendix of the manual technocratised

---

Fourth for the Amendment of the Law and the Improvement of the Administration of Justice in Criminal Cases (9 George IV c. 66, 1828). This Act adopted in New South Wales *An Act for Improving the Administration of Criminal Justice in England* (7 George IV c. 64, 1826) ss 4–6.

14 The third edition of MacNevin's manual compiled in Part IV of the Appendix a number of cases decided by the Supreme Court of New South Wales on the jurisdiction of the office of coroner. It also contained a collection of relevant English Law Reports, which was noticeably absent from the second edition.

15 Government Gazette (Sydney), No 22, 15 March, 1845, 300.

the role of the coroner. The term 'technocracy' was coined by William Henry Smyth in the early twentieth century to describe the work of government through knowledge, expertise and skill.[16] I employ the term here to describe how the conduct of the coroner was transformed in the late nineteenth century by the dissemination of technical knowledge in professional manuals. The anthologisation of circulars, as well as precedents and forms, emphasised that technocratic processes were important for coroners, if not essential for the fulfilment of their obligations of office. Or, to put this differently, anthologisation proffered technocratic activities, such as form-filling, writing letters and book-keeping, as practices immanent to the efficacy of the death investigation process. Through the manual, these technologies were represented as properly belonging to the institutional life of coronial law.

It must be stressed that the function of precedent as presented in the coronial manual was not the same as its operation in the common law. Precedent did not only concern the binding of a case to the declaration of the laws of the colony. Rather, in a more expansive way, it involved the attachment of administrative procedure to the obligations of office. Precedent consisted of template proclamations to be issued at the opening of an inquest or for the making of oaths or affirmations.[17] It also assumed form in the instructions requiring coroners to provide specific information on inquest forms,[18] and transcribe and transmit the forms according to exact guidelines. Witness depositions, for instance, were to be recorded on 'half-sheets of foolscap paper, quarter margin, and only on one side of the sheet'; while a copy of the inquest proceedings, including vouchers, were to be circulated to the Department of Justice '*under registered cover* ... with duty stamps affixed'.[19] The validity of an inquest depended not

---

16 William Henry Smyth, *Technocracy Part III: Ways and Means to Gain Industrial Democracy* (Gazette, 1920). See further on a history of governance by expertise, Nikolas Rose, 'Expertise and the Government of Conduct' (1994) 14 *Studies of Law, Politics and Society* 359.

17 For example, the coroner was instructed to direct the constable to open the court by proclaiming

> You good men of this district, summoned to appear here this day to inquire for our Sovereign Lady the Queen when, how, and by what means A.B. came to his death, answer to your names as you should be called, every man at the first call, upon the pain and peril that shall fall thereon.
>
> MacNevin, *Manual for Coroners and Magistrates in New South Wales*, 3rd edn, above n 1, 34.

18 Coroners were required to complete the Form of Inquisition (Form 5) and divide the form into three parts: 'the *caption*, the *finding*, and the *attestation*': Ibid, 55 (emphases in original).

19 Ibid, 41 and 57 (emphasis in original). The Department of Justice issued a circular on 16 November 1891 requesting that coroners use registered cover to transmit inquest proceedings due to the 'inconvenience' of a number of cases 'having gone astray in transit to this Department': Ibid, 92. For an example of this in other Australian colonies, see *Acts Relating to Coroners and Instructions for the Guidance of Coroners on Justices Acting As Such*, above n

only on whether the dead body was viewed by the coroner (and the jury), as discussed at length in the previous chapter, but also on whether the inquest documents were properly 'fastened with tape or a metal fastener'.[20] The absence of a seal accompanying documents could lead to delays, if not a declaration that the inquest was void.[21] The inconsistency in the procedure followed in adjoining districts and the irregularities of transcription and transmission of inquest proceedings contributed to a growing need to standardise the conduct of coroners in Australia. Indeed, such problems were cited by MacNevin in the Preface to the second edition of his manual as one of the reasons why he decided to write a handbook for the coroners of the Colony of New South Wales.

In compiling precedents, forms and circulars, certificates, warrants and vouchers, the professional manual technocratised what was often considered in the first half of the nineteenth century to be a highly disorganised, unprofessional and parochial office. But another way of theorising the impact of the manual on the conduct of office is that it theatricalised the way in which coroners performed their duties. It set out precise instructions for legal actors and, at times, provided a script book for the ceremonial staging of law:

> The Coroner then asks them if they have agreed in their verdict, and if they say '*Yes,*' he asks them, '*Who shall say for you?*' to which they will say, '*Our foreman.*' The Coroner will then say, '*Mr. Foreman: How do you find A.D. came to his death, and by what means?*' The foreman then standing, relates and gives the verdict, which it is the Coroner's duty to receive, enter, and record.[22]

Stage cues for the conduct of inquests, which were meticulously outlined in manuals, promoted a range of rhetorical devices for authorising the lawfulness of the coroner. Inquests could only begin once the coroner directed the constable to proclaim the opening of the court: 'You good men of this district, summoned to appear here'.[23]

10, 24 (emphasis in original): 'The inquisitions, depositions, recognisances, and coroner's remarks (if any), should be carefully annexed together with tape and seal, and forwarded by the *first mail* to the Crown Law Offices; exact attention to this is indispensable'.

20 MacNevin, *Manual for Coroners and Magistrates in New South Wales*, 3rd edn, above n 1, 41.

21 Though it is to be noted that, according to William Ramsay Smith, then coroner of Adelaide,

> [n]o inquisition is to be quashed on account of technical defects; and if exception be taken to any inquisition on account of such, any judge of the Supreme Court may, if he thinks fit, order the inquisition to be amended.
> William Ramsay Smith, *A Manual for Coroners: Being a Guide to Coronial Inquiries and Inquests in South Australia and throughout Australasia and in England* (Hussey & Gillingham, 1904) 72.

22 MacNevin, *Manual for Coroners and Magistrates in New South Wales*, 3rd edn, above n 1, 51 (emphases in original).

23 Ibid, 34.

What was also of particular importance for framing the lawfulness of the coroner was a collection of forms for remunerating individuals who took part in inquest proceedings. Coroners were required to apply for an advance from the Department of Justice to pay jurors for attending an inquest, compensating them for their time spent on the jury, and to submit accounts of expenditure, 'properly vouched and receipted',[24] to the Audit Office within a fortnight of the date on which funds were advanced. Here, the persona of the coroner – what Dorsett and McVeigh define as 'the manifestation or expression of office – the way a person fills that office'[25] – was re-imagined through the character of the bureaucrat. Coroners performed this persona when requesting forms from the Government Printer, issuing vouchers, paying fees, writing legible reports and, notably, ensuring that all correspondence was properly sealed and stamped.

The coronial manual thus reshaped the technical operations of the death investigation process in the late nineteenth century. Eighteenth-century guidebooks had always included a limited number of forms that had to be completed by coroners in the conduct of office. However, subsequent iterations of this genre amplified the technocratic role of the coroner. The manual supplemented judicial statements on current law with circulars, precedent and forms, and augmented official instructions with opinions, advice and minutes from civil servants and government ministers. The language of technocracy was embedded in and through the manual.[26] While its inventory of circulars, precedents and forms undoubtedly technocratised the office of coroner, it is precisely through this logic that the manual held on to the question of an ethics of responsibility.

The bone collector and the coronial textbook

William Ramsay Smith was one of the most questionable physicians to have ever assumed the office of coroner in Australia. Smith was educated in arts, natural sciences and medicine at the University of Edinburgh from 1877 to 1892. He was first employed as a physician at the Royal Adelaide Hospital in 1896 and then appointed as the Inspector of Anatomy, Chairman of the Central Board of

---

24 Ibid, 10.
25 Shaunnagh Dorsett and Shaun McVeigh, 'The *Persona* of the Jurist in Salmond's *Jurisprudence*: On the Exposition of "What Law is ... "' (2007) 38 *Victoria University of Wellington Law Review* 771, 773.
26 Hilary Golder remarks that, by the end of the nineteenth century, the office of the coroner 'was fully integrated into the magisterial career structure': above n 12, 118. She explains that, following the enactment of the *Public Service Act 1895* (NSW),

> politicians ceded control of government employment to a central and independent personnel agency. The new Public Service Board was responsible for the classification, discipline and promotion of public servants, but above all for their recruitment by means of competitive or qualifying examination.
>
> Ibid, 119

Health and Coroner of Adelaide in the Colony of South Australia in 1899.[27] Smith first raised the ire of the medical profession by holding inquests into deaths in hospitals and lunatic asylums, particularly following surgery, which had not until then been conducted in the colony.[28] However, what caused the most controversy was his fondness for 'unlawfully' dissecting what he called 'government corpses' in his Adelaide morgue at West Terrace Cemetery.[29]

Smith maintained that the purpose of his 'experiments' was to test his theories of death causation. It was reported that he used mallets to cause puncture wounds, knives to create stab wounds and a range of bullets to analyse the distinctive effects of using different weapons on the flesh.[30] Yet the coroner was also dismembering (primarily Indigenous, though some non-Indigenous) corpses and sending their remains – including skulls, organs, skin and genitalia – to the University of Edinburgh and the Royal College of Surgeons in Edinburgh, or adding them to his own personal collection. Smith became renowned by the beginning of the twentieth century as a 'prolific colonial collector' of the remains of the Ngarrindjeri people, which were well documented in his numerous manuscripts on anatomy, anthropology and odontology.[31] He gleefully compared – in pamphlets co-written with Sir Edward Charles Stirling on Indigenous burial grounds – his adventures in grave robbing to prospecting for gold.[32]

Smith's unlawful experiments at the Adelaide morgue were made possible through the manipulation of legal rules.[33] In the late nineteenth century, coroners were permitted, after the end of an inquest, to dispose of an unclaimed

27 He wrote over 80 books, pamphlets and journal articles on a range of topics, including: William Ramsay Smith, *A Description of Some Tasmanian Skulls* (Australian Museum, 1900); William Ramsay Smith, *Medical Jurisprudence from the Judicial Standpoint* (Stevens and Sons, 1913); William Ramsay Smith, *In Southern Seas: Wanderings of a Naturalist* (John Murray, 1924). In the late twentieth century, it was argued that he plagiarised the writings of Indigenous author David Unaipon in *Myths and Legends of the Australian Aborigines* (Dover Publications, [1932] 2003 edn). In Smith's entry on 'The Aborigines of Australians' in the *Official Year Book of the Commonwealth of Australia* (Commonwealth Bureau of Census and Statistics, 1909), he wrote that Indigenous Australians '[are the] most interesting [race] at present on earth and the least deserving to be exterminated by us and the most wronged at our hands': Ronald Elmslie and Susan Nance, 'Smith, William Ramsay (1859–1937)' in *Australian Dictionary of Biography, Volume 11* (Melbourne University Press, 1988). http://adb.anu.edu.au/biography/smith-william-ramsay-8493
28 Helen MacDonald, *Possessing the Dead: The Artful Science of Anatomy* (Melbourne University Press, 2010) 190.
29 Ibid, 191. See also Helen MacDonald, 'The Anatomy Inspector and the Government Corpse' (2009) 6(2) *History Australia* 40.1, 40.1.
30 Claire Scobie, 'The Return of Bones' (2009) 68(4) *Meanjin*. www.meanjin.com.au/editions/volume-68-number-4-2009/article/the-return-of-the-bones
31 Ibid.
32 William Ramsay Smith and Edward Charles Stirling, *Australian Aborigines: Burial Grounds* (Thomas Gill, 1907–1911).
33 MacDonald, above n 28, 194.

corpse – defined as a corpse that was not claimed by relatives within 12 hours of death – by ordering its burial or sending it to a medical school if the dead had not objected to such use of their cadaver prior to their passing. Smith side-stepped these rules by first issuing a warrant for an unclaimed corpse's burial following an inquest, and then secretly 'filleting' the corpse prior to placing what remained of its cadaver in a wooden coffin.[34] His unlawful conduct was exposed in 1903, when he was charged with being 'indiscreet' in his dealings with the dead under the decency clause of the *Anatomy Act 1884* (SA) ('*Anatomy Act*').[35] He claimed, though, before the Board of Inquiry that was set up in the wake of the scandal that he was

> unsure whether a coroner had that power to dispose of unclaimed bodies when an inquest concluded and spoke of having asked the government to change the law to enable him to do so since inquest corpses often made excellent subjects for dissection.[36]

While Smith was initially suspended from the office of coroner once his conduct was exposed, the Board of Inquiry eventually cleared him of any wrongdoing and commended him for 'his painstaking research' in medical jurisprudence.[37] In 1904, he was reinstated to this office – but not to the role of Inspector of Anatomy, which was found to conflict with his other duties – and he continued his practice of dissecting, dismembering and collecting Indigenous corpses. In the same year, Smith wrote *A Manual for Coroners*, which not only set out guidelines for holding inquests in South Australia, but also compared the legal framework in that state to coronial laws across Australia, New Zealand and England.[38]

---

34 Ibid.
35 Claire Scobie summarises the specific case that catalysed the scandal:

> [The Board of Inquiry] centred on how Smith handled the body of Tommy Walker, a popular Aboriginal figure who lived on the streets of Adelaide. On Walker's death in July 1901, city businessmen were so moved they paid for a carved headstone and two obituaries ran in the local papers. But within hours of his death, Smith had intercepted, cut up and decapitated Walker's body. The coffin, containing only his flesh, was weighed down with sand. [When] news travelled back from Edinburgh that Walker's remains now formed part of its Anatomy Museum [the] revelations triggered public outrage.

> above n 30.

36 MacDonald, above n 28, 194.
37 Elmslie and Nance, above n 27. This seems to be in line with Smith's biography in the *Australian Dictionary of Biography*, where Ronald Elmslie and Susan Nance characterise him as a 'courageous, conscientious, effective and diligent public servant' who 'administered his duties without fear, favour or rancour'.
38 The colonies of New South Wales, Queensland, Tasmania, Victoria and Western Australia formed a Federation of States of Australia in 1901.

A Manual for Coroners was innovative in a number of ways for how it intervened in the genre of the professional handbook. In the first instance, it transformed the manual into a textbook that standardised the art of anthologising
forms, circulars and precedent. The textbook was divided into seven parts, that
corresponded to different elements of the inquest proceeding. Each chapter in
Parts I to V began with an examination of coronial law and procedure in South
Australia, before highlighting the differences from and similarities to other states
and countries. Part VIII listed coronial legislation from England, New Zealand
and Australia. In contrast to MacNevin's manual, this edition did not contain an
appendix; instead, it integrated circulars, precedents and forms throughout its
comparative analysis of coronial law and procedure. The emergence of a new
genre of the textbook correlated with a pluralisation of types of readers. The
enlarged scope of the subject matter appealed to a more diverse audience,
including students, doctors, barristers, solicitors, justices and coroners, who
sought an authoritative guide for navigating the procedures of coronial institutions in different common law jurisdictions.

In the second instance, the textbook provided practical advice on how coroners should behave in their office. Following his exoneration by the Board of
Inquiry of all charges of indiscreet dealings with the dead, Smith took great care
to account for the manner in which he conducted his office. In Part VI, he set
out the 'Liabilities and Privileges of a Coroner', and reiterated a passage from
MacNevin's manual that a coroner was not liable for any acts committed by him
'in his judicial capacity, and *within scope of his jurisdiction*'.[39] However, he went
further by insisting that a coroner was free 'from all *vexatious* actions for any act
done by him in his judicial capacity'.[40] It is notable that Smith conspicuously
omitted from his manual any discussion of the possibility that coroners may
contravene the *Anatomy Act* for unlawfully dissecting the dead. On the other
hand, it is not surprising that he cautioned his readers that coroners may only
be fined or imprisoned under the common law for corruption in office, and
could *only* be removed from office for misbehaviour, wilful neglect or incapacity
in the discharge of a duty.[41] Indeed, this revealed much about how his previous
experiences before governmental inquiries had affected the way he conceptualised the privileges and liabilities of holding office.

Notwithstanding that Smith's advice for coroners seemed to suggest that any act
done by them in their judicial capacity was immune from accountability, the most
contentious aspect of Part VI proved to be his instructions on when a coroner
should decide to hold an inquest. Earlier in the textbook, Smith maintained that
inquests into deaths in hospitals and asylums, particularly following surgery – a

---

39 MacNevin, *Manual for Coroners and Magistrates in New South Wales*, 3rd edn, above n 1, 18
(emphasis in original).
40 Smith, *A Manual for Coroners*, above n 21, 94 (emphasis in original).
41 In the eighteenth century, coroners could only be removed from office if they were convicted
of extortion, wilful neglect of duty or a misdemeanour of office.

practice that, as I mentioned above, raised the chagrin of the medical profession – should fall under the scope of the coroner's jurisdiction. Coroners were bound to exercise their jurisdiction in hospitals and asylums because 'the public mind should be satisfied that persons in such institutions, who are outside the circle of their relatives and acquaintance, received proper medical care and attention'.[42] Yet he further explained that coroners were *obliged* to investigate such deaths not simply because the public demanded it, but because of the occupation of their office. The issue here was that the decision to hold an inquest did not involve a question of jurisdiction alone. The obligation to hold an inquest into a hospital or asylum death was imposed by a coroner's fealty to their office.

The problem of discretion was broader in coronial law than the vexed issue of whether to hold an inquest into a hospital or asylum death. Smith counselled that '[r]efusing without adequate reason to hold an inquest' or '[h]olding an inquest unnecessarily' constituted an offence that could result in liability.[43] The primary reason for this was that an inquest incurred a considerable expense for governments, particularly given that all jurors, witnesses and coroners were paid for their attendance during proceedings. In fact, most coroners throughout the nineteenth century were paid partially by salary and partially by a fee for conducting an inquest.[44] They were also reimbursed for any expenses incurred while travelling to the place of inquest. Unnecessary proceedings were viewed by government officials with suspicion as an entrepreneurial activity, or at worst as a distraction from the kind of unlawful actions of which Smith was accused before the Board of Inquiry.

Given the personal stakes at play, Smith sought to advise coroners on precisely when they should exercise their discretion to hold an inquest. The usual course of action was for coroners to make a decision once they had received a report of the death from a constable. According to English case law, coroners may only assume jurisdiction over a dead body after they have received an official notice of the death. Smith quoted from *R v Price* in support of this supposition:

> The coroner has no absolute right to hold inquests in every case in which he chooses to do so. It would be intolerable if he had power to intrude without adequate cause upon the privacy of a family in distress, and to interfere with the arrangements for a funeral. Nothing can justify such interference, except a reasonable suspicion that there may have been something peculiar in the death, that it may have been due to other causes than common illness. In such cases the coroner not only may, but ought to hold an inquest ... He should not go, generally speaking, until he is sent for.[45]

---

42 Smith, *A Manual for Coroners*, above n 21, 6.
43 Ibid, 93. See also MacNevin, *Manual for Coroners and Magistrates in New South Wales*, 3rd edn, above n 1, 15.
44 Ian Freckleton and David Ranson, *Death Investigation and the Coroner's Inquest* (Oxford University Press, 2006) 48.
45 *R v Price* (1884) 12 QBD 247, 248.

However, he also surmised that 'he would not be acting illegally if he proceeded on his own knowledge or upon the information of a private person' – that is, alongside 'he honestly believes [the] information which has been given to him to be true, which, if true, would make it his *duty* to hold such inquest'.[46] In short, the problem of discretion was ultimately a question of duty. While the 'truth' of the information received about a death rather than the way it was obtained was an important determinant of the decision to hold an inquest, what is most interesting about this passage is the emphasis Smith placed on ethics. In quoting the judgment of *Stephenson*, yet reframing it as his own opinion of the law, Smith asserted that in certain situations coroners were compelled by fealty to their office to investigate the cause of a death. Even if a coroner's intentions were unlawful, Smith's manual advised them that, upon the receipt of 'truthful' knowledge indicating a sudden, violent or unnatural death, they were obliged by their office to hold an inquest into that death.

Part VI of Smith's textbook demonstrates how the genre of the professional handbook offered coroners a model of ethical conduct in the early twentieth century. It advised coroners on when, where and how they should fulfil the obligations that pertained to their office. Their conduct was undoubtedly constrained by laws, norms and customs; but, despite these constraints – which in the case of Smith seemed hardly effectual – performing the role of coroner required discretion, and interpretation of what responsibilities pertained to the role and how they should be exercised. The conduct of the coroner's office thus required the development of an ethical mindset. For Jeffrey Minson, this involved acting in accordance with 'duties that are incumbent on a person by virtue of his or her occupancy of a particular role or position', while rejecting 'self-interest, partisan interest or factional interest'.[47]

Cultivating an ethics of office

The language of *office*, which derives from the Latin word *officium* and the Greek word *leitourgia*, refers to the conduct of a role. The Roman jurist Cicero described *officium*, which he translated from the Greek word *kathekon*, in *De Officiis* as 'what is appropriate, opportune' in the circumstances.[48] *Officium* was not equivalent in Ancient Rome to 'a juridical or moral obligation nor a pure and simple natural necessity'.[49] Rather, it determined what one ought to do and how one ought to behave in a social situation – that is, what was most

---

46 Smith, *A Manual for Coroners*, above n 21, 13 and 93. He quoted on page 93 the English case of *R v Stephenson* (1884) 13 QBD 331, 331 (emphasis added).
47 Jeffrey Minson, 'Holding on to Office', in David Burchell and Andrew Leigh (eds), *The Prince's New Clothes: Why Do Australians Dislike Their Politicians?* (UNSW Press, 2002) 128 and 137.
48 Giorgio Agamben, *Opus Dei: An Archaeology of Duty* (Adam Kotsko trans, Stanford University Press, 2013) 67.
49 Ibid, 72.

appropriate to do in the circumstances in which one occupied a particular role. To appear in public or private life, in households or in the agora, every (free male) person occupied a role, an *officium*, and during their incumbency they were bound by certain powers, rights and privileges. By the seventeenth century, the presupposition of office denoted duty, obligation and responsibility, whether moral or juridical:

> an office was an identifiable and discriminate constellation of responsibilities and subordinate rights, or liberties asserted to be necessary for their fulfilment, manifested in a *persona* and regarded as in some way socially necessary or acceptable.[50]

The vocabulary of office continued to exert influence on law and politics in the eighteenth and nineteenth centuries, despite the ascendancy of a discourse of rights.[51] One of the reasons for this is that the term 'office' was used to designate the conduct of government, in particular the organisation of the public service. The state constituted a network, a multiplicity of offices, while public servants became typified by the office they assumed. Office talk referred to

> an institution that the state and other juristic bodies of public law make use of in order to accomplish certain purposes. Sovereign and fiscal tasks are delegated to a persona – the 'office-holder' – for a portfolio of responsibilities that is delimited, amongst other things, by norms of competence. These persons – state functionaries or bureaucrats – are subject to official duties that result, *interalia*, from legislation, constitutional dictat or official instructions.[52]

This means office talk came to define the way the state delegated a portfolio of responsibilities to office-holders, as well as the manner by which these office-holders fulfilled their obligations. The notion that an office was performed indicates that office-holders had to cultivate a persona – one encompassing technical skills, expertise and competencies – to carry out the duties that pertained to that specific role.

Coronial manuals offered guidance on the cultivation of a persona of office in sections that set out processes for the appointment of coroners. In the eighteenth century, the manual explicitly linked the problem of who should occupy the office to the question of character. The criteria for assuming office in

50 Conal Condren, *Argument and Authority in Early Modern England: The Presupposition of Oaths and Offices* (Cambridge University Press, 2006) 29.
51 Condren claims that '[t]he fear of incommensurable offices simultaneously held was a spur to Kant's destruction of office by treating ethical activity as only one universal world of duty': Ibid, 28.
52 Paul du Gay, *Organizing Identity* (Sage Publications, 2007) 106.

England were described in relation to the qualities of a person – there were five in total – which were designed to 'prevent Men of small Value and little Under-standing to be chosen Coroners'.[53] He was required to be a good man (*Probus Homo*), a lawful man (*Legalis Homo*), knowledgeable, capable and diligent. He was also required to be a landowner such that he could exercise his duties with-out drawing a salary. The criteria for becoming a coroner were transformed in Australia. Not only were coroners in Australia appointed directly by colonial governments – whereas in England, county coroners were elected officials and occupied the role for life until 1887[54] – and generally provided with a salary, but there were also no special criteria for being appointed to the office. That is alongside the fact that they could demonstrate 'in a general way' knowledge of the coronial jurisdiction:

> it is presumed that gentlemen accepting the office, and taking upon them-selves the duties, are acquainted in a general way with those duties, and pre-pared to act on their own independent judgment and discretion in all matters in relation thereto in such manner as they may think best in the public interests and in the interests of the administration of justice.[55]

The absence of specific criteria for entrance to the coroner's office was lamented by William F.A. Boys in his introduction to *A Practical Treatise on the Office and Duties of Coroners in Upper Canada* (1864):

> Formerly the office of coroner was of such high repute that no one under the degree of knighthood could aspire to its attainment ... It has, however, now fallen from such pristine dignity and though still of great respectability, no qualifications are required beyond being a male of the full age of 21 years, of sound mind, and a subject of her Majesty, and possessing the amount of education and mental ability necessary for the proper discharge of the duties.[56]

---

53 *The Coroner's Guide*, above n 8, 4.
54 Ian Burney, *Bodies of Evidence: Medicine and the Politics of the English Inquest, 1830–1926* (Johns Hopkins University Press, 2000) 3. The *Coroners Act 1887* (UK) transformed the eligibility requirements for becoming a coroner, while the *County Coroners Act 1860* (UK) provided coroners with a salary. While the majority of coroners in Australia were medically qualified, English coroners were mostly legally trained. For further information, see Ian Burney's discussion of Thomas Wakley's successful election as the medical coroner for Middlesex in 1830 in Chapter 1.
55 MacNevin, *Manual for Coroners and Magistrates in New South Wales*, 3rd edn, above n 1, 2.
56 William F.A. Boys, *A Practical Treatise on the Office and Duties of Coroners in Upper Canada* (1864) 2, quoted in Myles Leslie, 'Reforming the Coroner: Death Investigation Manuals in Ontario 1863–1894' (2008) 100(2) *Ontario History* 221, 226. Most coroners in the United States of America possessed no qualifications in medicine or law. They were 'occupants of a minor political office, and were typically farmers, carters or undertakers': Freckleton and Ranson, above n 44, 71. The political nature of the role led to the demise of the coronership

Boys emphasised in subsequent editions the importance of developing a certain persona – namely, characterised by literacy, refinement, chivalry, *noblesse oblige* and sociality – for gaining appointment as a coroner in British Canada. He advised aspiring coroners to seek a 'recommendation of a member of Parliament, or other person possessing influence with the Executive'.[57] He even counselled them to show vigour when seeking office.[58] Myles Leslie claims that Boys's manual clearly professionalised the office of coroner in Canada. It 'reformed [the office] as a professional instrument of modern government, rather than as a minor parochial office'.[59] While this is not to be disputed, Boys's emphasis on developing an apropos persona for assuming the office of coroner achieved much more than this. It revealed how the technology of the manual held on to the question of an ethics of responsibility. It demonstrated that when coroners occupy their office they must assume a persona, and through the expression of that persona, they must cultivate an ethics of office.

The question of how to form an ethics of office was also shaped by the difficulties of negotiating any potential conflict between the different personae an individual could occupy in civil society. Like other coronial manuals at the turn of the twentieth century, Smith's textbook advised coroners not to act in cases where they may have attended the dead 'professionally during the last illness or at the time of the death of such person' or acted as solicitors 'in the prosecution or defence of a person for an offence for which such person is charged by an inquisition taken before him as a coroner'.[60] However, what was of greater significance was his advice against acting in office while assuming other roles within the civil service. The South Australian Board of Inquiry acquitted Smith in 1903 of all charges that he had contravened the *Anatomy Act* in dissecting, filleting and collecting Indigenous and non-Indigenous remains. But it also found that there was a conflict of interest between the multiple roles he occupied. Smith was exonerated before the inquiry on the condition that he must choose to assume the role of either coroner or inspector of anatomy. His textbook thus advised coroners that an ethical mindset towards the dead necessitated at the very least mitigating conflicts that could potentially arise from adopting multiple personae in civil society.

It is important to remember that coronial manuals were written in the aftermath of the Northcote-Trevelyan reforms in Britain, and their gradual implementation in British colonies. While the Northcote-Trevelyan Report was

in certain states of America in the early twentieth century and its substitution with a system of medical examination. The latter was stripped of the former's judicial powers and appointed as a public servant. See further, Robert H. Vickers, *The Powers and Duties of Police Officers and Coroners* (T.H. Flood, 1889).

57 Ibid, 227.
58 Ibid.
59 Ibid, 225.
60 Smith, *A Manual for Coroners*, above n 21, 1; MacNevin, *Manual for Coroners and Magistrates in New South Wales*, 3rd edn, above n 1, 20.

published in 1854, as Thomas Osborne writes, its recommendations 'were not properly addressed until Gladstone's reforms of the 1870s (and even then not properly realized)'.[61] The report recommended widespread changes in the conduct of government: reductions in public expenditure, centralisation of auditing processes, technocratisation of the public service, and increased independence and autonomy for office-holders. Its publication has been remarked by historians as a significant turning point in the reform of the British civil service in the nineteenth century. It signalled a desire 'to fabricate administration as an autonomous ethos or art, separated both from the pull of political patronage and from narrow, specialized expertise'.[62] Bearing in mind the trajectory of these reforms, I will outline in the remaining pages of this chapter how the technology of the manual played an integral role in holding coroners accountable for their conduct. The coronial manual addressed the question of an ethics of responsibility by framing the death investigation process within a bureaucratic logic of office.

In Max Weber's monumental opus on economics, politics and sociology, he claims that hierarchy, supervision, subordination and centralisation were all characteristics of Western bureaucracies in the nineteenth and twentieth centuries. He describes bureaucracy in Volume 3 of *Economy and Society* as the institutional pursuit of rational administration through the use of technical knowledge, skill and expertise. Bureaucratic administration promoted a logic of objectivity, efficiency and continuity by eliminating from the activity of decision-making 'love, hatred, and all purely personal, irrational, and emotional elements which escape calculation'.[63] This logic posited that every decision and judgement was conditioned by 'a system of rationally debatable "reasons" ... either subsumption under norms, or a weighing of ends and means'.[64]

The development of a modern bureaucracy in British colonies involved the implementation of Northcote-Trevelyan's reforms of government. It also involved a transition from honorific service to salaried officials, and the substitution of tenure for life with the appointment of office-holders for a limited period of time. The reliance upon a money economy conditioned the possibility of bureaucratic administration and positioned it as technically superior to other forms of governance, such as administration by patronage. The bureaucratic form also prohibited the use of personal discretion in favour of rational decision-making. It demanded from office-holders an impersonal, functional and unbiased interpretation and application of law, and a strict adherence to the procedural conditions of legal institutions. Yet this demand was only ever an ideal,

---

61 Thomas Osborne, 'Bureaucracy as a Vocation: Governmentality and Administration in Nineteenth-Century Britain' (1994) 7(3) *Journal of Historical Sociology* 289, 293. See further, Oliver MacDonagh, 'The Nineteenth-Century Revolution in Government: A Reappraisal' (1958) 1(1) *The Historical Journal* 52.

62 Ibid, 294.

63 Max Weber, *Economy and Society: An Outline of Interpretive Sociology* (Guenther Roth and Claus Wittich trans, Bedminster Press, 1968) 975.

64 Ibid, 979.

for it failed to consign office-holders, as Weber writes, to the role of 'an automaton into which legal documents and fees are stuffed at the top in order that it may spill forth the verdict at the bottom along with the reasons, read mechanically from codified paragraphs'.[65] Weber resolutely rejects this image of the conduct of law for it implies an iron-clad 'bureaucratization of justice'.[66]

Many scholars have mischaracterised Max Weber's depiction of bureaucracy in *Economy and Society* as an 'iron cage' devoid of ethics. They have often misinterpreted 'bureaucratic administration', as du Gay notes, as 'the result of the disintegration of morally unified personhood'.[67] Indeed, Weber's writings on bureaucracy are much more nuanced than they have been made out to be. He associates bureaucracy with 'the plural creation of historically specific ethics or Lebensführungen (instituted conducts of ethical life)'.[68] In other words, Weber denounces the notion that bureaucracy was ever inconsistent with an ethics of responsibility. And he does so by emphasising that '[e]ntrance into an office ... is considered an acceptance of a specific *duty of fealty* to the purpose of the office'.[69] Weber uses the language of fealty to explain how bureaucrats occupied an 'order of life' (*Lebensführung*), how they expressed themselves through an 'instituted persona' and how the performance of this persona involved the cultivation of a specific ethics of office.[70]

In his writings on Weber, *In Praise of Bureaucracy*, du Gay categorises fealty as the 'ethos of bureaucratic office' – that is, 'a certain ethical dignity for a particular form of institution – the bureau – and the category of person – the bureaucrat'.[71] He reproaches critiques that reductively position conscience against procedure, ethics against rationality, and conduct against technocracy. For Weber, the bureau was an 'order of life', which developed its own distinctive ethical conduct, and bureaucrats conducted themselves according to the formation of an ethical mindset that pertained to this life-order (*Lebensführung*).[72] A bureaucratic mindset then, such as 'strict adherence to procedure, acceptance of sub- and superordination, abnegation of personal moral enthusiasms, [and] commitment to the purposes of the office',[73] constituted the ethical substance of civil office at the turn of the twentieth century. This substance was 'the product of particular ethical techniques and practices through which individuals

---

65 Ibid.
66 Ibid.
67 Paul du Gay, *In Praise of Bureaucracy: Weber, Organization, Ethics* (Sage Publications, 2000) 29.
68 Paul du Gay, 'Max Weber and the Moral Economy of Office' (2008) 1(2) *Journal of Cultural Economy* 129, 134.
69 Weber, *Economy and Society*, above n 63, 959 (emphasis added).
70 du Gay, 'Max Weber and the Moral Economy of Office', above n 68, 134–135.
71 du Gay, *In Praise of Bureaucracy*, above n 67, ix.
72 Jeffrey Minson, 'Bureaucratic Culture and the Management of Sexual Harassment' (Cultural Policy Studies, Occasional Paper No. 12, Institute for Cultural Studies, Division of Humanities, Griffith University, 1991) 13.
73 du Gay, *In Praise of Bureaucracy*, above n 67, 29

develop the disposition and capacity to conduct themselves according to the ethos of bureaucratic office'.[74]

Weber's writings on bureaucracy in *Economy and Society*, published posthumously in Germany in 1922, provide a context for understanding his main concerns which were presented in a series of lectures he delivered on the concept of vocation in 1919, shortly before his death. Weber argues in 'Politics as a Vocation' that bureaucratic office emerged as a vocation in the nineteenth century. Here, vocation (*Beruf*) means a calling, a training and a profession. To occupy an office is thus to answer a call, to commit oneself to its purpose, to devote oneself to its demands. Vocation shapes the formation of an ethical character: it defines the role of the bureaucrat in terms of delegated responsibilities and frames the way that bureaucrats ought to conduct themselves in office. It demands, as Weber wrote, 'a highly developed sense of professional *honor* with an emphasis on probity'.[75] The call that all bureaucrats had to answer was therefore that of fealty to office. This was a call to dutifully fulfil the obligations that pertained to the *occupation* of an office.

Coroners were not just any kind of bureaucrats, however. They had unique demands placed upon them by virtue of their vocational devotion to administrating a death investigation process. The responsibilities of their office first and foremost obliged them to form an ethical mindset towards the dead. This meant exercising discretion as to whether to hold an inquest into a hospital or asylum death, for example, and mitigating any conflicts caused by assuming different personae in civil society. It also meant transcribing verdicts on half-sheets of foolscap paper, sealing these documents with tape and sending them under registered cover with duty stamps to ensure that all inquests were validated. Coroners were obliged by fealty to their office to fulfil specific duties – even the most mundane procedural duties associated with a culture of form-filling – that had as their ultimate aim the attachment of the dead to the institutional life of coronial law. The professional manual in turn advised coroners on how best to meet the ethical demands of their office.

It could be claimed that coroners were far from responsible at the turn of the twentieth century, as exemplified by Smith's unlawful experiments on dead bodies in the Adelaide morgue. Or, rather, in light of revelations of how he disposed of unclaimed corpses upon the completion of an inquest, it could be suggested that the cultivation of an ethics of office was a poor description of how Smith actually occupied the role of coroner. While I do not disagree with this assessment of Smith's conduct, it does misconstrue how technologies of office, such as the coronial manual, transformed the notion of an ethics of responsibility.

74 Ibid, 4.
75 Max Weber, *The Vocation Lectures* (Rodney Livingstone trans, Hackett Publishing Company, 2004) 44 (emphasis in original).

Scott Veitch comments that legal institutions organise irresponsibility by 'rely[ing] on their legal obligation as a way of *evading* the very question of responsibility for their decisions and their consequences'.[76] He further notes that

> the obligations – and the fulfilment of the obligations – of the task, office or status are circumscribed accordingly. 'Acting responsibly' therefore means acting according to these obligations. When it comes to role responsibility then, responsible action means action within the permissible range, as established by the obligations of the role, making it irresponsible not so to act. ... failure to fulfil these obligations may lead to legitimate censure, but it will be censure *only* in accordance with that role.[77]

In other words, it was through performing his legal obligations, in acting upon a fealty to office, that Smith could dispense with any wider notions that he owed a social or moral responsibility to care for the remains of the Indigenous dead. This is perhaps best exemplified by his claim before the Board of Inquiry that he thought he was lawfully permitted as coroner to dispose of an unclaimed corpse in a manner of his choosing following an inquest. It is also denoted by claims made in coronial manuals that coroners were not liable for any acts, particularly 'vexatious actions', committed by them in their 'judicial capacity, and *within scope of his jurisdiction*'.[78]

The transformations of the technology of the manual in the late nineteenth and early twentieth centuries were concomitant with changes to the vocation of the coroner. No longer honorary office-holders, but salaried employees of a government department, coroners became increasingly accountable for their conduct in office. While this transition was far from a complete and seamless process, and was undoubtedly influenced by the reshaping of the occupation of the English coroner from an elected freeholder to an appointed employee, it did reflect a broader trend towards a bureaucratic logic of the coroner's office.[79] And it was augmented by the reliance on technical knowledge in the conduct of inquests, the emergence of a culture of form-filling, the demand for strict adherence to precedent, the monitoring of coronial expenses, the auditing of inquest documents and the centralisation of decision-making. However, as I have shown in this chapter, the bureaucratic logic of the coroner's office was far from incompatible with the cultivation of an ethics of responsibility.

In the preface to *A Manual for Coroners*, Smith cautioned his readers that knowledge of legislation and the common law alone was insufficient for performing

---

76 Scott Veitch, *Law and Irresponsibility: On the Legitimation of Human Suffering* (Routledge-Cavendish, 2000) 21 (emphasis in original).

77 Ibid, 48 (emphasis in original).

78 MacNevin, *Manual for Coroners and Magistrates in New South Wales*, 3rd edn, above n 1, 18 (emphasis in original). See also Smith, *A Manual for Coroners*, above n 21, 94.

79 See Paul Finn, 'The Law and Officials', in R.A. Chapman (ed), *Ethics in Public Service* (Edinburgh University Press, 1993) 139.

coronial duties. 'It is as difficult to conduct an inquest properly from merely reading acts of parliament', he wrote, 'as it is to make a logically connected speech from perusing a dictionary'.[80] What was necessary was a guidebook that instructed coroners on the technical procedures of the jurisdiction, not only in Australia, but also New Zealand and England. In the twentieth century, the coronial manual became a comparative textbook that guided a pluralised audience through the bureaucratic logic of the coroner's office.

This chapter has offered a historical account of the transformations of the coronial manual in the late nineteenth and early twentieth centuries. It has demonstrated how this genre technocratised the way in which coroners conducted themselves in office, while at the same time holding on to the question of how to form an ethics of responsibility. The manual undoubtedly provided tools for bureaucratising what was often thought of as a highly inefficient, disorganised office. Upon the completion of an inquest, coroners were required to file, fasten and transmit documents, including jurors' and witnesses' paysheets, to government departments; submit warrants to bury corpses to the police; and tabulate and transmit particulars for annual publication in the Government Gazette. Yet the character of the bureaucrat was not antithetical to the duties of the coroner. The obligation to issue vouchers, submit forms, collect statistics and stamp correspondence formed part of the ethical mindset of the office. The manual set out this bureaucratic logic in minute detail, but so too, did it hold coroners accountable for the manner in which they performed their duties.

The technology of the manual reshaped how coroners formed legal relations with the dead. It bureaucratised the office of coroner while setting out the liabilities and privileges of holding this office, and the duties, obligations and responsibilities that pertained to the assumption of the role. This chapter has investigated what was involved in the shift towards a bureaucratic logic of the coroner's office, what that shift allowed to come into existence and how that shift was represented by changes to the technology of the manual. Dorsett and McVeigh write that '[m]uch of the administration of modern law takes place through bureaucratic modes of writing – an obvious example is the form'.[81] In an institutional history of coronial law, this *form*, the form of writing, the art of anthologising and the technical devices of the manual, bound the dead to the institutional life of coronial law. The manual revealed that coronial law could never simply be expressed in the interpretation of a case or even the proclamation of a verdict, but more broadly in how the coroner formed an ethical mindset towards the dead in the performance of their office.

80 Smith, *A Manual for Coroners*, above n 21, v.
81 Shaunnagh Dorsett and Shaun McVeigh, *Jurisdiction* (Routledge, 2012) 62.

4   Dead records

In 'Drawing Things Together', Bruno Latour points out that the rational character of bureaucracy is not to be located in the psyche of the bureaucrat. The civil servant does not hold a unique propensity for rationalisation, any more than the scientist enjoys a predilection for objectivity. Instead, the institutional pursuit of technocratic administration is to be found in files. Latour compares the bureau to a small laboratory organised by an economy, a cascade of paperwork. 'In our cultures "paper shuffling" is the source of an essential power', writes Latour, 'that constantly escapes attention since its materiality is ignored'.[1] In *The Making of Law*, Latour demonstrates how the materiality of files organises a court of law. Files are embedded in the passage of law in the *Conseil d'Etat*, such that '[e]very case, at least in our countries of written law, is physically enveloped in a carton folder, held together with elastic bands'.[2]

The previous chapter traced a bureaucratic logic in the transformations of the coronial manual and its inventory of circulars, forms and precedents in the late nineteenth and early twentieth centuries. Yet what was missing from that account was a sustained analysis of how records, files and documents, created, collected, submitted and reproduced in the performance of office, shaped the bureaucratic impulse of the coronial jurisdiction. This chapter examines how record-keeping became an integral part of the modernisation of the coroner's court at the turn of the twentieth century. It first argues that the conduct of office was institutionalised as a court of record, a sitting of the coroner's court, through the technology of the file. Second, it considers the effects of this technology both on the role of the coroner, who increasingly assumed responsibility for recording a biography of the dead, and on the dead, who appeared as neither things nor persons, but records in an archive of institutional memory. The chapter concludes by suggesting that

---

1 Bruno Latour, 'Drawing Things Together', in Michael Lynch and Steve Woolgar (eds), *Representation in Scientific Practice* (MIT Press, 1990) 55.
2 Bruno Latour, *The Making of Law: An Ethnography of the Conseil d'Etat* (Marina Brilman and Alain Pottage trans, Polity Press, 2010) 70–71.

the duty of record-keeping was essential to how coroners took care of the dead. Coroners collected biographical information about the deceased, narrating their lives and deaths, through the technology of the file, which has come to signify one of the most important functions of the coronial jurisdiction.

A court that keeps its records

The coronial inquest into the death of Joseph O'Callaghan conducted on 3 April 1899 was full of intrigue. Newspapers were awash with running commentaries about every salacious aspect of the case.[3] Joseph was an illegitimate child, whose biological mother had disappeared and biological father was unknown. He was found emaciated in the home of Catherine Dillon, a respectable nurse, who reported his death to police on the day that he died. Evidence was put forward during the inquest that Joseph died from artificial feeding, yet it was ambiguous as to whether his death was accidental or intentional. The coroner, Samuel Curtis Candler, suspected foul play when he discovered that Joseph had been 'adopted' by Mrs Dillon. The child's 'unknown' father had paid Mrs Dillon £90 to adopt his son and he required her to sign a deed of agreement, written by a solicitor, which stipulated that the child should never know the name or identity of his biological father.[4] When the coroner asked Mrs Dillon to name the father, she refused to answer. Her initial response was that he was already married and had a family and she was afraid that, by identifying him, she would tarnish his reputation. However, after the coroner repeatedly adjourned the inquest and threatened her with a warrant for contempt of court, Mrs Dillon explained that if she were to answer the coroner's question she might incriminate herself under the *Infant Life Protection Act 1890* (Vic).[5] The irate coroner, clearly unimpressed by her unwillingness to respond to his demands, issued a warrant committing her for contempt of court. If Mrs Dillon did not answer his question within 14 days, she would be sent to gaol until the contempt was purged.

This inquest is notable not simply because it exemplified the media's insatiable appetite for the macabre on the eve of the twentieth century. Much to the chagrin of the public and the press, who flocked to the inquest following the purge of the contempt in the hope of catching a glimpse of the mysterious father providing evidence, Candler declared that, in the interests of the family, the father's name would be suppressed from publication and the proceedings

---

3 See 'Traffic in Babies: The O'Callaghan Case – The Unknown Father', *The Argus* (Melbourne), 12 May 1899, 6; 'The O'Callaghan Case: Appeal to the Full Court', *The Argus* (Melbourne), 24 June 1899, 7; 'The O'Callaghan Case: Verdict of Death from Natural Causes', *The Argus* (Melbourne), 5 July 1899, 9.

4 'An Infant's Death', *The Argus* (Melbourne), 4 April 1899, 3.

5 'Solicitor's Privilege: The O'Callaghan Case – Opinion of the Attorney-General', *The Argus* (Melbourne), 10 May 1899, 10.

would be held 'in camera'.[6] The inquest of Joseph O'Callaghan is significant because it resurrected the problem of how to define the jurisdiction of the coroner's court. It troubled what was thought to be settled law by questioning whether coroners, by virtue of their office, have the authority to commit a witness for contempt of court. Lawyers for Mrs Dillon promptly appealed against Candler's decision and argued that the authority of the coroner was limited by statute.[7] The integral question before the law then was jurisdictional and its answer depended on an often forgotten aspect of the history of the common law: the doctrine of 'court of record'.

O'Callaghan's inquest was not the first time that Candler found himself in the unenviable position of having his authority challenged by a superior court. The inquest that catalysed the 1874 case of *Casey v Candler* involved the discovery of a dead body in the seaside town of Brighton. Candler, at the time the coroner for the district of Bourke, promptly attended the scene of death to hold an inquest upon view of the body and called forth Dr C.G. Casey, who performed the post-mortem examination on the corpse, to give evidence of what he had found. There is little information about what exactly piqued Candler, but he became increasingly irritated by the doctor's performance on the stand and 'directed the plaintiff, in a peremptory manner, to describe the appearances in a different order'.[8] *The Argus* reported that the stoic physician rejected the interference of the coroner and refused to follow his direction. He said that 'he would give his evidence as he thought right'.[9] This led to an 'altercation', which culminated in the coroner committing Casey for contempt of court and directing a constable to escort him to the local gaol until the contempt was purged. In the aftermath of the squabble, Casey brought an action against the coroner for assault and false imprisonment.

Casey v Candler questioned whether a coroner could lawfully commit a person for contempt of court. It was argued by both the plaintiff and the defendant that the writ of contempt required judicial power and a coroner could only possess this power if the inquest could be said to constitute a court of record. In other words, the key issues in this case were whether coroners

---

6 'The O'Callaghan Case: Proceedings in Private – Father's Name Suppressed', *The Argus* (Melbourne), 28 June 1899, 9; 'The O'Callaghan Case: The Agreement Withheld from the Press', *The Argus* (Melbourne), 3 July 1899, 6. The conduct of the coroner was denounced at the time in the Parliament of Victoria. It was alleged that Candler's 'prurient' pursuit of the name of the unknown father gave air to the dangerous circulation of gossip:

> The action of the coroner had caused a very disagreeable feeling in the city, for statements found their way into circulation which reflected most unjustly on well-known citizens. It gave pain to several prominent persons ... Justice had been brought into contempt.
> 'The Coroner's Powers', *The Argus* (Melbourne), 6 July 1899, 7.

7 *In Re O'Callaghan* (1899) 104 VLR 957. For example, by arguing that the *Coroners Act 1890* (Vic) s 4 did not grant him any powers to commit a witness for contempt.

8 'Law Report', *The Argus* (Melbourne), 28 November 1974, 4.

9 Ibid.

possessed judicial or ministerial powers, whether the inquest was a mere enquiry or a court of law and, if it was the latter, whether the conduct of an inquest instituted a court of record. The plaintiff answered all three questions in the negative:

> A court of record is a court which *keeps its records*; the coroner has to send his proceedings to the Crown Law Offices; the proceedings of the justices are filed in the court of General Sessions. The coroner does not hold a court of record, nor is he a judicial officer ... There may be certain matters in England in which the coroner holds a court and acts judicially, but an inquest *super visum corporis* is not one of them. He merely holds the enquiry, reduces the result to writing, and sends it in to the law officers of the Crown.[10]

The coroner does not sit on a court, let alone a court of record, according to this line of argument, because the procedure for submitting files following an inquest to the Crown Law Offices severs the relationship between the act of writing and the act of recording. The inscription of speech, involving the transmission of voice to a text, is not enough for a legal officer to hold a court of record; as the plaintiff argued, '[t]he filing in some court of record, is essential to the constitution of a record'.[11]

It should be of little surprise that Justice Barry rejected the claims of the plaintiff and held that the coroner's court was in fact a court of record. The judge drew upon precedent in England as well as other British colonies to declare that the coroner had always been a judge of record.[12] The coroner was one of the most honourable officers of the Crown, not a lowly civil servant exercising limited ministerial powers; when sitting as a coroner on an inquest, their stature had to be respected as much as that of any judge. Their authority still resided in the common law, even though the office was regulated by statute, and the Antipodean coroner thus inherited all of the powers that the English coroner possessed. Even if one were to question whether the English coroner

---

10 *Casey v Candler* [1874] 5 AJR 358, 360 (emphasis added).

11 Ibid, 361.

12 Barry J cites *Garnett v Ferrand* [1827] 6 B&C 611, *Thomas v Churton* [1862] 2 B&S 475 and *R v White* (1860) 121 ER 394. The English case of *Garnett v Ferrand* [1827] 6 B&C 611, 612 involved a journalist attending an inquest 'for the purpose of wrongfully and illegally taking notes of and publishing the proceedings of the said inquisition'. The legal issue at stake here was whether the journalist could maintain an action for trespass against the coroner, who directed a constable to throw him out of the room for his actions. Lord Tenterden CJ maintained that 'no action will lie against a Judge of Record for any matter done by him in the exercise of his judicial functions': Ibid, 625. Hence, what was important in this case was the meaning of an open court. If the coroner's court is truly open, then the coroner should have the same powers as a judge of record, namely, the power of exclusion. See also a case that was not mentioned by Barry J, but which took place prior in the Colony of New South Wales: *Chippett v Thompson* (1868) 7 SCR (NSW) L 349.

holds a court of record, Barry J reminded all parties that a court of record was one in which all proceedings were recorded in writing. It was not necessary for the records to be filed, but simply that they be recorded.[13] The coroner had jurisdiction to commit a witness for contempt of court by virtue of their office, which bestowed upon them the judicial authority of a judge of record.

Reaffirming the decision of *Casey v Candler*, Chief Justice Madden in *Re O'Callaghan* held that, under the common law, the coroner's court was indisputably a court of record.[14] He stressed that his decision was based upon a survey of precedent from England. In the Middle Ages, the coroner held one of the most important offices in England and, while modern legislation had gradually whittled away its authority, it nonetheless commanded the power to 'enforce the conduct of its own proceedings with propriety and safety'.[15] He further explained that it was 'the very nomination of a person to sit as coroner [that] carried with it all the rights of the coroner's court'.[16] Despite the prominence given to the word 'right' in this sentence, what stands out in Madden CJ's judgment is his emphasis on the question of conduct. For what was at stake in determining whether a coroner sits in a court of record was not whether they possessed the right to make inquiries or issue a writ of contempt, but whether they had authority to *conduct* their office prudently. This means that the legal determination that the coroner was a judge of record endowed them with a panoply of technologies to manage their court as they saw fit.

Parchment, rolls and files

The technology of the file was undoubtedly intrinsic to the jurisdictional question of whether the coroner's court was one of record. This is most apparent by tracing the history of the doctrine in the Middle Ages. Court of record originally referred to the 'indisputability' of oral accounts and the minuting of proceedings on 'parchment roll'.[17] The enrolment of decisions in the twelfth and thirteenth centuries marked a court as one of record and set forth a practice of judgment based on what had been previously inscribed on parchment. In other words, written records conditioned the possibility of the doctrine of precedent and denoted that any decisions made in writing were conclusive. It is important to note, though, that the roll did not immediately supersede oral accounts of proceedings. Instead, the idea was propagated that writing was more definitive

---

13 Barry J also made the somewhat tautological argument that a court of record is established by the coroner's powers, as set out in legislation, to fine or imprison. He declared that 'no court can fine or imprison which is not a court of record' and 'where there is a power *de novo* created by Parliament to fine and imprison, either of these two makes it a court of record': *Casey v Candler* [1874] 5 AJR 358, 363.

14 In *Re O'Callaghan* (1899) 104 VLR 957, 963.

15 Ibid, 964.

16 Ibid.

17 Samuel Edmund Thorne, *Essays in Legal History* (Hambledon Press, 1985) 254.

and more accurate than collective memory. 'Records were acts and judicial proceedings enrolled on parchment', Enid Campbell writes, the truth of which was
'not to be called into question'.[18]

The official title of the coroner in the twelfth and thirteenth centuries was
Custos Placitorum Corone, which meant 'keeper of the pleas of the Crown'.[19]
The original duty of the medieval coroner was to keep the pleas of the Crown
by recording legal decisions on parchment rolls:

> The function implied by their title is that of keeping (*custodire*) as distin-
> guished from that of holding (*tenere*) the pleas of the crown; they are not
> to hear and determine causes, but to keep record of all that goes on in the
> county and concerns the administration of criminal justice, and more par-
> ticularly must they guard the revenues which will come to the king if such
> justice be duly done.[20]

The distinction between keeping and holding pleas – the former duty belonged
to the coroner, while the latter pertained to the justice of the peace – was codified
in Chapter 24 of the *Magna Carta* in 1215. The *Magna Carta* effectively depicted
the coroner as an itinerant annalist, a state-sponsored archivist, travelling across the
lands of England, transcribing the proceedings of courts of law and detailing with
precision, on rolls of parchment, revenue collected on behalf of the Crown.

Coronial rolls were essential to the administrative practice of keeping the
pleas. Inquests on dead bodies were to be chronicled in 'files of small pieces of
parchment, each containing the record of a single case in note form'.[21] The coroner's fiscal duties – including the collection of fines, amercements and forfeitures, and revenue from treasure troves, royal fish and the valuation of deodands –
would have been arduous, if not impossible, without recording in writing the
proceedings of different courts throughout the country. However, what should
be emphasised here is that the coroner was tasked with not only transcribing
oral accounts into written records, but also *keeping* them. That is, they were

---

18 Enid Campbell, 'Inferior and Superior Courts and Courts of Record' (1997) 6 *Journal of
Judicial Administration* 249, 256. The emergence of written records in the common law has
been said to derive from the Norman conquest of 1066 and the use of the Domesday Book;
however, 'written law' has a much longer lineage in Roman jurisprudence: see M.T. Clanchy,
From Memory to Written Record: England 1066–1307 (Blackwell, 2nd edn, 1993).
19 R.F. Hunnisett, *The Medieval Coroner* (Cambridge University Press, 1961) 1. Hunnisett
notes that the official title was 'continued to be used throughout the Middle Ages'. The
shortened title of *coronator* (coroner or crowner) gradually took hold from the thirteenth
century onwards.
20 Sir Frederick Pollock and Frederic William Maitland, *The History of English Law before the
Time of Edward I* (Cambridge University Press, 1898) 534. See also F.J. Waldo, 'The Ancient
Office of Coroner' (1910–1911) 8 *Transactions of the Medico-Legal Society* 101, 108; R.F.
Hunnisett, 'The Origins of the Office of Coroner' (1958) 3 *Royal Historical Society Transactions* 85.
21 Hunnisett, above n 19, 114.

required to copy files onto larger pieces of parchment and store that parchment in formal rolls, which could be presented to another court in the future – particularly when a superior court, such as the Eyres, arrived in town and demanded that coroners show it their rolls. R.F. Hunnisett notes that the practice of keeping records was integral to the responsibilities of the coroner, such that '[t]he first duty known to have been performed by any coroners was the personal recording of an appeal of homicide by the Lincolnshire coroners in the *Curia Regis* in November 1194'.[22] It was not until the sixteenth century that coroners were not only keeping rolls of files, but also submitting *duplicates* of those files to other offices of the Crown.

In the centuries that followed the Middle Ages, particularly after the invention of the printing press in the fifteenth century, the doctrine of court of record came to denote the development of specific legal procedures inherent in a court that preserved its decisions in writing. For example, only a court that kept records of its judgments, according to Sir Edward Coke, could fine or imprison a person for contempt of court.[23] While the doctrine initially referred, then, to the recording of speech in writing, by the nineteenth century it had become a referent of a complex hierarchy between inferior and superior courts of law, where the latter were defined by their capacity to issue a writ of contempt. '[T]he writ is no ordinary writing or warrant, no simple missive of everyday provenance', Peter Goodrich writes, 'but is rather a heavy sign replete with the full panoply of institutional obscurities, opaque lexical insignia, repetitions and the boredom of complex formal detail that attend to legal address'.[24] The writ of contempt, as it appeared in the doctrine of record, referred not only to a literacy of law, a writing that emerged before speech, but to law as record, to the files that made the passage of law possible.

This history of the doctrine of court of record gestures towards a constitutive relationship between the institutional life of law and the technology of the file. For Weber, the emergence of bureaucratic governance in the nineteenth and twentieth centuries could only take place through the production, collection and dissemination of files.[25] Cornelia Vismann, however, makes a more specific historical argument in *Files: Law and Media Technology*: that record-keeping is a foundational condition of the institutionalisation of law. She writes that law is 'a repository of forms of authoritarian and administrative acts that assume concrete shape in files'.[26] To put this differently, law and files are mutually constitutive.

---

22 Ibid, 98–99.
23 Campbell, above n 18, 255.
24 Peter Goodrich, 'Visive Powers: Colours, Trees and Genres of Jurisdiction' (2008) 2(2) *Law and Humanities* 213, 214.
25 Max Weber, *Economy and Society: An Outline of Interpretive Sociology* (Guenther Roth and Claus Wittich trans, Bedminster Press, 1968) 957.
26 Cornelia Vismann, *Files: Law and Media Technology* (Geoffrey Winthrop-Young trans, Stanford University Press, 2008) xiii. In this text, Vismann draws a distinction between the materiality of files and the discursivity of records, which is not reflected in the German word for 'file': *Akten*.

The former assumes its institutional form in the recording of its proceedings, yet files acquire their materiality in the institutional practices of law. Vismann contends that a historiography of legal institutions must first consider the role that files played in mediating relations between orality and writing, which were said to have marked the emergence of the common law during the Norman conquest of England.

> According to this binary, the validity and security of the law belong to writing, whereas lack of duration, individual case law, and greater proximity to specific events are associated with orality. The evolution of the law finds its place in between the two, presenting itself as a history of progressive rationalization and intensified literacy.[27]

In other words, Vismann rejects the teleological narrative that law was born out of 'the substitution of writing for orality'.[28] But at the same time she suggests that the origins of law do not lie in 'an oral culture', but rather in 'administrative record keeping'.[29] In this regard, files are 'one of the oldest of the legal profession's technologies', Patricia Tuitt argues, and '[t]he tabulation and storage of the minutiae of human existence is one of its most enduring and important routines'.[30]

Files are irreducible to discourses of literacy or orality. They neither conform to rules of phonetic writing nor those of rhetorical speech. Their force in law, which Vismann compares to that of the list, is conditioned by their presence within a writing–speaking dichotomy. Indeed, files are constitutive elements of legal institutions because they function 'in quasi-oral fashion'; that is, they 'shed light on their own development, on that which precedes grand acts of writing such as the clean copy or that legal institutes and institutions'.[31] The production of files reveals the formation of administrative procedures of law; they show how legal institutions congeal and jurisdictional technologies make ordinary actions lawful: 'The unlimited capacity for addition and circulation turns files into a medium of presence. It endows them with the same characteristics as speech, with the result that they appear to be up-to-date, live, ever-changing, acting, and inexhaustible'.[32]

This is not to say that files lie outside the 'medium of writing'. Far from it: Vismann contends that 'files capture everything that other forms of writing no

---

This chapter deploys the terms 'files' and 'records' interchangeably to acknowledge their entanglement in the institutional practices of coronial law.

27 Ibid, 3.
28 Ibid, 4.
29 Ibid.
30 Patricia Tuitt, 'Legal Practice and Modes of Dying: Bruno Latour, Technology and the Critical Legal Instance' (2005) 16 *Law and Critique* 113, 122.
31 Vismann, above n 26, 8.
32 Ibid, 10.

longer contain – all the life, the struggles and speeches that surround decisions'.[33] They circulate in what Michel de Certeau calls the 'scriptural economy', a set of operations by which speech acts conceal themselves in writing as a trace, mark or remainder of what lies outside the text.[34] '*The place from which one speaks is outside the scriptural enterprise*', de Certeau opines, '[t]he uttering occurs outside the places in which systems of statements are composed'.[35] The strategies of the scriptural economy are translation and transcription, which attempt to make sense of speech, and transform orality into textual narratives. Files are firmly entrenched in this economy of operations, since they inscribe speech acts in forms of writing, while also containing that which haunts writing.

Files thus assume their shape through the institutional operations of the scriptural economy. They are created through bureaucratic processes and administrative procedures and, as Latour has shown, are embedded in the making of laws. To this point, this chapter has suggested that record-keeping played a significant role in the institutionalisation of the coroner's court at the turn of the twentieth century. The transformation of the office of coroner into a court of law was only made possible through the technology of the file. This is further exemplified by a parliamentary debate that took place in the early twentieth century about the meaning of the word 'inquisition'.[36] The debate centred around the syntactic difference between an 'inquest' and an 'inquisition'. Most legislators sought to define the latter as either a document, a report or writing, whereas a minority sought to delineate it as an inquiry, a judgment or a finding. In other words, there was disagreement as to whether an inquisition was scriptural or procedural. While one politician argued that the word denoted 'writing that is made after or during the inquest', or a decision, finding or verdict *in writing*, another retorted that the phrase 'an inquisition in writing' was so ambiguous such that 'it implies that there may be an inquisition which is not in writing'.[37]

What was at stake here, on the one hand, was a question of hermeneutics. How does one distinguish the speech acts of coroners from their written judgments? What are the rhetorical means by which a voice becomes a record of a decision? On the other hand, the debate problematised the jurisdiction of the coroner. Does only the record of an inquest authorise the decision of the coroner? In debating the meaning of the word 'inquisition', legislators emphasised how technologies of writing, as well as those of filing and recording, were integral to the conduct of an *inquest* and, more broadly, an essential aspect of the institutional life of coronial law. They highlighted that the real issue at stake was

---

33 Ibid.
34 Michel de Certeau, *The Practice of Everyday Life* (Steven Rendall trans, University of California Press, 1984) 155.
35 Ibid, 158 (emphasis in original).
36 Victoria, *Parliamentary Debates*, Legislative Assembly, 19 October 1911. The debate concerned the enactment of the *Coroners Act 1911* (Vic), which repealed the *Coroners Act 1890* (Vic), *Coroners Act 1896* (Vic) and *Coroners Act 1903* (Vic).
37 Ibid, 2004 (Mr Mackay) and 2006 (Mr Prendergast).

not whether an inquisition was a document or a judgment, but rather how the scriptural became procedural and the procedural became scriptural. If the document 'descends from the Latin root *docer*, to teach or show', this means that the inquisition signified a document not simply because it recorded speech in writing, but because it was a bureaucratic action in itself.[38]

By the early twentieth century, both Parliament and the judiciary had come to recognise that the technical processes of filing were intractably attached to the jurisdiction of coroners. As I discussed above, a court of record was first and foremost a technology of record-keeping, and only subsequently became a jurisdictional device to manage questions of authority. In case law and parliamentary debates, technologies of filing, recording and writing clearly came to be conceived of as intrinsic to the institutionalisation of the coroner's office as a sitting of a court of law. The practices of transcribing the speech acts of coroners, and of writing and submitting inquisitions and other documents to bureaucratic departments, were equally (if not more) important for the institutional life of coronial law than the perennial question of whether the authority to commit a witness for contempt of court lay in common law or statute. The performance of the duties of the coroner's office was formalised as a court of record through the technology of the file. And the materiality of the file endowed coroners with the authority to hold a court, which bestowed powers upon them to conduct their court as they saw fit.

Making a case of the dead

In Part Three of *Discipline and Punish*, Foucault describes the 'examination' as a discursive technique particular to the transformation of penal institutions in the eighteenth and nineteenth centuries. The examination made use of 'an observing hierarchy' and 'a normalizing judgment' to qualify, classify and punish prisoners in their individuality.[39] Yet, for the prisoner to be judged as an individual, more was required of the examination than the technology of the gaze. Individualisation involved describing and inscribing prisoners, and recording their actions, capacities and desires, which created 'a whole meticulous archive constituted in terms of bodies and days'.[40] 'Lives of a few lines or a few pages', Foucault writes elsewhere, 'nameless misfortunes and adventures gathered into a handful of words'.[41] The deployment of the examination as an institutional practice of record-keeping demonstrated how files were key to the

---

38 Lisa Gitelman, *Paper Knowledge: Toward a Media History of Documents* (Duke University Press, 2014) 1. On the circulation of documents in modern society, see Annelise Riles (ed), *Documents: Artifacts of Modern Knowledge* (University of Michigan Press, 2006).

39 Michel Foucault, *Discipline and Punish: The Birth of the Prison* (Alan Sheridan trans, Penguin Books, 1991) 184.

40 Ibid, 189.

41 Michel Foucault, 'The Lives of Infamous Men', in James D. Faubion (ed), *Power: Essential Works of Foucault 1954–1984, Volume 3* (Robert Hurley trans, Penguin Books, 2002) 157.

individualising process. They facilitated the creation of taxonomies, the discovery of patterns and the comparison of trends, which conditioned the prisoner as a describable, calculable entity in scientific discourse:

> The examination that places individuals in a field of surveillance also situates them in a network of writing; it engages them in a whole mass of documents that capture and fix them. The procedures of examination were accompanied at the same time by a system of intense registration and of documentary accumulation. A 'power of writing' was constituted as an essential part in the mechanisms of discipline. On many points, it was modelled on the traditional methods of administrative documentation, though with particular techniques and important innovations. Some concerned methods of identification, signalling or description.[42]

The examination was a record-keeping technology par excellence. It embedded bodies in networks of writing, it wrote bodies into records and files, which had as their model eighteenth-century paradigms of bureaucratic paperwork. In constructing the individual as an object to be studied, described and measured, the human sciences, including law, integrated this technology into all modes of governance. As Foucault notes, '[t]hese small techniques of notation, of registration, of constituting files, of arranging facts in columns and tables that are so familiar to us now, were of decisive importance in the epistemological "thaw" of the sciences of the individual'.[43] It is thus not possible to give a historical account of the appearance of the individual as an object of knowledge in the eighteenth and nineteenth centuries without also examining the history of files. The accumulation of records in archives, the storage, indexing and organisation of files in administrative cabinets, was imperative for ensuring the efficacy of the examination, the production of knowledge on the individual and the operation of a disciplinary power that was inextricable from the birth of the prison and other judicial institutions.

It is notable that a correlative effect of the use of the examination in penal institutions was the emergence of the 'case'. '*The examination, surrounded by all its documentary techniques, makes each individual a "case"*', Foucault observes,

> [and] clearly indicates the appearance of a new modality of power in which each individual receives as his status his own individuality, and in which he is linked by his status to the features, the measurements, the gaps, the 'marks' that characterize him and make him a 'case'.[44]

Cases were not equivalent to files. While the former were undoubtedly contained in the latter, their institutional functioning was to be found in how they

---

42 Foucault, above n 39, 189.
43 Ibid, 190–191.
44 Ibid, 191–192 (emphasis in original).

transformed discrete factual units into biographical narratives. Prior to the deployment of the examination as an institutional practice of record-keeping, Foucault claims, biographies were only written of 'privileged' men. Or, to put this differently, only a few 'famous' individuals had their lives chronicled, recorded and preserved for posterity prior to the eighteenth century. With the invention of the 'case', biographical accounts became a fate that everyone endured, however ordinary, lowly or ignoble an individual was in life. What was once a rarefied genre of literature, extolling the virtues of a life lived well, the ancestral biography was replaced by technocratic writing, which stripped it of its literary flourishes, reducing the individual's life to a set of mundane, comparative characterisations.

This fate did not only befall the 'dangerous individual', who was the sole focus of Foucault's analysis in *Discipline and Punish*; it also had consequences for the dead who were subject to a coronial inquest. The inquisition, the document tendered by coroners at the conclusion of an inquest, consisted of three parts: a caption, a verdict and an attestation. The caption listed the name of the deceased, while the verdict specified the cause of death or, more specifically, the time, place and circumstance of death. Classification systems for death causation were initially based in British colonies on William Farr's nosological index, but were later modified to adapt to local maladies, such as sunstroke and snake bites. In the twentieth century, the body of the inquisition was expanded to touch upon the evidence of witnesses, the life of the deceased, and how the way they lived came to affect their death. Other records produced by coroners, such as certificates of death, which listed the age of the deceased, place of birth, location of death and cause of death, and notices for the Registrar of Births, Deaths and Marriages, which added the rank or profession of the deceased, created a trove of records of the dead.[45] The biographical information collated here was potentially illimitable: 'Christian name and surname of father? Christian name and maiden surname of mother? If deceased was married? To whom? Date? Where born? Undertaker's name? Any other matter of detail that may be valuable to Registrar'.[46]

45 It should be noted that, from 1836, legislation for registering births, deaths and marriages in England

> required coroners to see that their juries inquired into the particulars required by the new death certificates, and then to report their findings to the Registrar of the district in which the inquest was held (after 1874, this had to be done within five days).
> Olive Anderson, *Suicide in Victorian and Edwardian England* (Clarendon Press, 1987) 30.

46 Thomas E. MacNevin, *Manual for Coroners and Magistrates in New South Wales: Being a Practical Guide to the Proceedings of the Coroner's Court and to the Holding of Magisterial Inquiries in Lieu of Inquests by Justices of the Peace* (Charles Potter, Government Printer, 3rd edn, 1895) 119.

The coroner's *examination* of the dead created a genre of institutional writing that properly belonged to law. In contrast to the literary genre of the obituary, which was like the ancestral biography, reserved for the privileged few, coronial record-keeping produced the most mundane biographies, terse tributes to lives tragically cut short and prosaic narratives of the causes of a death. The coroner's records made the dead into a 'case'. They individualised the dead by recording biographical information, organising that information into 'comparative fields' and, when taken in their entirety, gathering together an institutional, collective memory of the dead, which could be disseminated to newspapers, registrars, statisticians and physicians. Where I depart from Foucault is in arguing that the case was never solely 'a means of control and a method of domination'.[47] It was also a tool for engaging lawfully with the dead, and a jurisdictional technology that attached a biography of the dead to the institutional life of coronial law:

> Every coroner was obliged to deliver his account in the shape of a register of inquests held and expenses incurred to the paying authority every four months and to draw up at the end of every inquest the formal legal document embodying the jury's finding known as an inquisition; but whether or not he kept any other records, and what form they took, was entirely for him to decide.[48]

The modality of this institutional attachment was archival; the dead only became files when they were stored, indexed and named in an archive.[49] This description of the conduct of the coroner recalls Vismann's account of Melville's Bartleby – 'a recording machine that without any sign of fatigue copies whatever it is presented with'.[50] Yet, as I have remarked previously, this image of bureaucratic administration was only ever an ideal. The inquest could not proceed, as I discussed in Chapter 2, without a *quality* view of the corpse; if the files were improperly fastened, stapled or couriered, as seen in Chapter 3, the inquest was deemed void and the record of the dead was cancelled. The incomplete file or inadequate record-keeping also became a problem for establishing the jurisdiction of the coroner's court, as I demonstrated earlier in this chapter. The first effect of the technology of the file on the institutional life of coronial law was that it transformed the dead – who appeared in law as neither things nor persons – into a case, a record, a file.

---

47 Foucault, above n 39, 191.
48 Anderson, above n 45, 37. Anderson discusses the differences between the record-keeping practices of several nineteenth- and twentieth-century coroners, such as Serjeant W.J. Payne, who attached 'hastily scribbled notes of the evidence given' in his inquisitions, and F.J. Waldo, who bound his papers into three volumes: 'first, the inquisitions themselves; then the depositions and other case papers; and finally a "No-inquest" series': Ibid, 38.
49 The archive is for discourse '*the system of its enunciability* ... [and] *its functioning*': Michel Foucault, *The Archaeology of Knowledge & The Discourse on Language* (A.M. Sheridan Smith trans, Pantheon Books, 1972) 129 (emphasis in original).
50 Vismann, above n 26, 31.

The question that remains is how did this record-keeping technology affect the conduct of the coroner's office at the turn of the twentieth century? Earlier in this chapter, I demonstrated that a duty to record derived from the history of the medieval coroner. The obligation to keep records, transcribe documents and file inquisitions in rolls of parchment was inherent in the performance of the office in the Middle Ages. It was a central component of the duties of the medieval coroner and to this extent typified the legal relations that the coroner established with the dead. Yet, following the emergence of the case in the eighteenth century, this duty took on additional significance. The duty to record arose from files themselves, not simply from the performance of office. One way of considering this shift has been by tracking the transformations of the coronial manual, but another way is by writing a history of the birth of the archive and examining how the archive developed an ethics of the file. In concluding this chapter, I assert that the archive imposed an obligation on coroners to narrate a biography of the dead, and to record the particulars of their lives and their deaths.

In the wake of the French Revolution, the modern archive was born.[51] The National Archive of France was established in 1790, while the Public Record Office of England was founded in 1838.[52] Public Record Offices were established throughout Australia in the twentieth century. In this sense, I am defining the archive as a governmental institution for the accommodation, serialisation and organisation of official records, files and documents. For Mike Featherstone, '[t]he archive was part of the apparatus of social rule and regulation, it facilitated the governance of the territory and population through accumulated information'.[53] Influenced by Foucault's analysis of the case as a technique of disciplinary power, Featherstone considers the archive as 'the place in which the sacred texts and objects were stored that were used to generate collective identity and social solidarity'.[54] In the nineteenth century, the archive came to signify a citadel of the history of the nation-state.

It also became a site through which the living could encounter a collective memory of the dead. Memory assumes form through shared discourses of language, techniques of narration and a social framework of living together.[55] Or, as Pierre Nora writes, 'memory is above all archival. It relies entirely on the materiality of the trace'.[56] *Les lieux de mémoire* mark the traces of memory in different institutions, such as cemeteries, museums, monuments, courthouses

51 For a fascinating history on the life of paperwork during and after the French Revolution, see Ben Kafka, *The Demon of Writing: Powers and Failures of Paperwork* (Zone Books, 2012).
52 Mike Featherstone, 'Archive' (2006) 23(2–3) *Theory, Culture & Society* 591, 592.
53 Ibid, 591.
54 Ibid, 592.
55 Maurice Halbwachs, *On Collective Memory* (Lewis A. Coser trans, The University of Chicago Press, 1992) 175.
56 Pierre Nora, 'Between Memory and History: *Les Lieux de Mémoire*' (Marc Roudebush trans, 1989) 26 *Representations* 7, 13. See also Thomas Osborne, 'The Ordinariness of the Archive' (1999) 12(2) *History of the Human Sciences* 51.

and hospitals. All of these places institute encounters between the past and the present, the dead and the living, the collective and the individual. The archive emerged in the eighteenth and nineteenth centuries as another *lieu* for embracing a collective memory of the dead. The administrative organisation of this institution, the way traces of the dead were attached to the materiality of files, functioned 'to block the work of forgetting, to establish a state of things, to immortalize death, to materialize the immaterial'.[57]

These institutional arrangements ushered in 'a radical new ethics of paperwork, one designed to sustain a state whose legitimacy was founded on the claim to represent, at every moment, every member of the nation'.[58] It also did much more than this. The archive was not simply directed outwards towards strategies of governance. It was channelled inwardly towards the ethical conduct of officials who produced copious amounts of paperwork that found their final resting place in archives. Indeed, techniques of filing not only transformed individuals into cases; they documented the process of documentation, recorded techniques of writing and transformed the disciplinary state, as Kafka writes, into a 'disciplined state'. The invention of the archive shaped the modalities of occupying an office. In the assumption of a role, all undertakings were to be recorded; every decision, action and task was to be indexed in writing. The performance of official duties was to be noted, inscribed and catalogued. No obligation was fulfilled until documents were preserved, submitted and received, and all of these actions were to be recorded in writing. As Kafka remarks, '[e]very action or transaction undertaken by any person with or on behalf of the state had to be documented in certain anticipation of an eventual accounting'.[59]

This new bureaucratic ethos attached itself to the institutional life of coronial law in the late nineteenth century. It referenced the original duty of the medieval coroner to keep pleas of the Crown, record decisions on parchment rolls and store copies of those rolls in files, which could be shown at will to other offices of the realm. Yet this new archival ethics also augmented those duties by demanding of coroners that they be held accountable for their record-keeping practices. The coroner became responsible for determining what kind of biographical information gleaned from the deceased should be recorded in inquisitions, what kind of records should be sent to the Registrar and other officeholders, how to classify the panoply of information collated during a death investigation, and how to store all of these records, files and documents in archives. By attaching the technology of the file to questions of conduct, the coroner shaped an ethics for taking care of records of the dead. The coroner formed an ethical mindset towards the coronial file, and assumed a responsibility of care through the act of recording – that is, by making a case of the dead. The technology of the file presented the dead as a case-narrative, a case-biography.

57 Ibid, 19.
58 Kafka, above n 51, 21.
59 Ibid, 44.

The appearance of a new archival ethics at the turn of the twentieth century thus meant that the coroner must assume responsibility for writing a biography of the dead; recording their name, age, profession and other particulars; and binding all of these discrete facts to the life of the coronial institution:

> Filing is a form of burying. The technological components of the file call to mind all the associated paraphernalia of the dead and of all forms of passing. ... When a story is narrated and reduced to an administrative file, something of its narrator remains; and it is something absolutely vulnerable because it has passed into another form and is already nearing its end. It is for these reasons that the keeper of the file bears a great responsibility. All of mortality is brought to life by the action of inserting a file in its allocated space in a filing cabinet, of the turn of the lock and of the regular routing out of files that have lain too long.[60]

Embalmed in a citation

The question of whether a corpse constitutes a person or a thing has long perplexed the common law. The seventeenth-century *Haynes Case*, which questioned whether a winding sheet could be stolen from a grave, defined the corpse as neither a person nor a thing (*res nullius*), 'but a lump of earth hath no capacity'.[61] Yet, since then, there have been more conflicting cases, such as the decision in *Doodeward v Spence*,[62] which declared that a corpse could only become a thing if, by the labour or skill of the living, it was transformed into something else. This chapter has suggested that the question of whether the corpse constitutes a person or a thing in law is unhelpful for tracing how coroners formed legal relations with the dead in the nineteenth and twentieth centuries. It is far more useful to conceive of the dead as records and files, and to consider how, by making them into cases, coroners bound their legacies to the everyday practices of the coronial institution.

The technology of the file transformed the way in which coroners took care of the dead in the late nineteenth and early twentieth centuries. The effects of this transformation are most clearly evident in the aftermath of the *O'Callaghan* case. While the jurisdictional question of whether the coroner's court was one of record was debated before the judiciary, the broader question of how law's files make the dead into a case, and the ethical implications of this jurisdictional device, was not lost on the public. In a popular magazine, *Table Talk*, William Fink published in 1899 a poem about the Baby O'Callaghan case, concluding that:

---

60 Tuitt, above n 30, 114.
61 (1614) 77 ER 1389, 1389.
62 [1908] HCA 45.

Though dead, has wrought himself a deathless name.
Embalmed in V.L.R and A.L.T,
Peeping, it may be, from his angel loft,
The while keen lawyers 'turn him up' to cite,
And quote him as of great authority.
Or e'er the child was in his leading strings
He is a leading case. The herd may rage
About his unnamed father, but the son
Though dead, yet speaketh, his unhappy case
Becomes a happy one, and Coroners
Sitting on hapless children, haply may
Do right to millions for all time to come.[63]

While the name of the baby's father was ultimately not recorded in the inquisition, much to the chagrin of a titillated press, what this poem reveals is the vital role that record-keeping assumed in making the dead into a case. Techniques of filing separated the name, Joseph O'Callaghan, from a corpse, and immortalised it in a citation – a legal reference that would remain integral to the development of the doctrine of coronial law in Australia. Even though Joseph's corpse was buried in the ground, far from the prying eyes of the legal fraternity, and his file gathered dust, forgotten in a cabinet, library or office, his name lived on, its presence haunting the courtroom, embalmed in a citation. This was for the coronial institution perhaps what Robert Pogue Harrison calls the 'afterimage' of the corpse. The history of this afterimage, a history of rolls, files and records, references an institutional shift in the conduct of the coroner's office at the end of the nineteenth century.

Embalming a name in a citation demonstrates a peculiar affinity with what Michel de Certeau writes about law in *The Practice of Everyday Life*:

> the law constantly writes itself on bodies. It engraves itself on parchments made from the skin of its subjects. It articulates them in a juridical corpus. It makes its book out of them. These writings carry out two complementary operations through them, living beings are 'packed into a text' (in the sense that products are canned or padded), transformed into signifiers of rules (a sort of 'intextuation') and, on the other hand, the reason or *Logos* of a society 'becomes flesh' (an incarnation).[64]

I have shown in this chapter that coroners had at their disposal an inventory of record-keeping technologies for transforming the corpse into a word, 'a single

63 William Fink, 'The O'Callaghan Baby Case [From the "Australian Law Times"]', *Table Talk* (Melbourne), 25 August 1899, 3.
64 de Certeau, above n 34, 140.

name that can be read and quoted by others'.[65] The earlier discussion of the history of the doctrine of court of record revealed how these technologies were integral to the formalisation of the coroner's office as a sitting of a court of law. I later examined how the dead became records, stored in an archive of institutional memory, while at the same time coroners, expressing the persona of archivist, assumed responsibility for curating those files. The file functioned as a jurisdictional device for attaching a biography of the dead to the institutional life of coronial law.

The image of an expansive coronial archive – a trove of files, records and documents; and annals of cases, citations and references – is evoked by Fink's poem. But the poem also laments that, when a case is made of the dead, when the dead are embalmed in citations, they become separated from their names. While their names live on in the history of law-making, chronicles of their lives remain forgotten. Fink thus reminds his reader that coroners, lawyers and judges all have a responsibility not only to remember the names of the dead, but also to take care of the material traces of their lives. And this responsibility can only be fulfilled by recording biographical information about the dead. Joseph O'Callaghan called Coroner Candler from his file. This call constituted an ethical demand to narrate his biography, to inscribe on paper the legacy of his short life. The call, which originated deep within the ethics of office, demanded that the coroner assume responsibility for curating an archive of dead records.

65 Ibid, 149.

5   Screening the corpse

In a short article written for *The Juridical Review* in 1896, Sir Henry Littlejohn, the prominent Scottish forensic scientist, opined that '[t]o render a medical report of a post-mortem examination as distinct and as easily comprehended as possible, is no easy matter'.[1] While his grievance was ostensibly directed towards the authors of such reports, as he was a forensic expert himself, it appeared more as a defence to charges levelled by the judiciary that post-mortem reports were inscrutable. Littlejohn acknowledged that it was surely difficult, if not a formidable task, to translate the wounds of the corpse into a language that could be understood by a legally trained judge. His solution to this quandary was the increased use of photography during the death investigation process. He recommended that '[e]very county constable, in my opinion, should be provided with a camera, by means of which he could indicate with unerring accuracy the attitude of a dead body'.[2]

During the year before, Wilhelm Conrad Röntgen discovered a new kind of 'photography', which, by the mid-twentieth century, had transfigured the practice of coronial law and forensic medicine. In 'On a New Kind of Rays', which was originally published in German in 1895 but translated a year later for *Science*, Röntgen demonstrated his discovery of what he called X-rays, a consequence of combining an induction coil with a cathode vacuum tube. The tube produced a fluorescence, which, when shone onto an object, rendered it transparent. One of the first experiments Röntgen conducted was on his wife's hand. The radiating light penetrated her skin, projecting 'shadows' of the bone onto a fluorescent screen, 'with only faint outlines of the surrounding tissues'.[3] He also experimented with capturing these impressions on a photographic plate and he named the images caused by the radiating light 'shadow photographs'. Röntgen

---

1 Henry Littlejohn, 'Photography and Criminal Inquiries' (1896) VIII *The Juridical Review: A Journal of Legal and Political Science* 13, 13.
2 Ibid, 16.
3 Wilhelm Conrad Röntgen, 'On a New Kind of Rays' (1896) 3(59) *Science* 227, 227. The article is a translation of *Sitzungsberichte der Würzberger Physik-medic. Gesellschaft*, which was originally published in *Nature* on 5 December 1895. It was translated into English by Arthur Stanton.

was not sure what constituted these new kind of rays, but he surmised that they could have been produced by longitudinal vibrations. Nor could he predict the profound effect that this optical apparatus would have on the disciplines of medicine and law in the twentieth century.

The final chapter of this book examines how forensic radiography transformed the way in which coroners took care of the dead in the twentieth century. The use of X-rays during the death investigation process altered the coroner's view of the corpse, a view that remained necessary in the late nineteenth century for an inquest to proceed. By the mid-twentieth century, the coronial jury ceased to view the corpse during an inquest, while the post-mortem examination was left almost entirely in the hands of the forensic pathologist. The invention of radiography 'mechanised' the forensic gaze that evolved in the nineteenth century; medico-legal representations of the corpse no longer relied upon the cutting up of bodies. The coroner's view of the corpse thus became increasingly mediated by medical imaging technology. Far from interpreting the mechanised gaze as one that detached the corpse from the institutional life of coronial law, this chapter argues that it transformed the quality of caring practices. Coroners had to develop a new aptitude for viewing the interiority of the corpse from afar, and acquire new skills for interpreting the significance of shadows as evidence of death causation. The history of forensic radiography demonstrates how new technologies continue to make demands on coronial institutions to think differently about how they cultivate legal relations with the dead.

Shadows of a corpse

News of Röntgen's discovery spread rapidly across the world in 1896. It was discussed at length in academic journals and press media, and even entered popular discourse in much of the West. The most interesting aspect of the imaging technology, which stunned both the medical and legal worlds, was that the shadows cast by X-rays, which when shone on the human body exposed densities lying underneath the skin, could be captured permanently on a photographic plate. These 'shadow records' were not ephemeral, which disrupted legal understandings of the admissibility of evidence in the same way that photography challenged the temporality of oral testimony in the late nineteenth century.[4] X-ray images were first admitted as evidence in civil cases in 1896.[5] By 1897, they were accepted as proof of gunshot wounds or other injuries in criminal trials.[6] The American case of *Smith v Grant*, which involved an action for negligence against a physician, was significant insofar as it marked the first time a court explicitly embraced medical imaging

---

4 For a historical account of the acceptance of photographic evidence in law, see Jennifer L. Mnookin, 'The Image of Truth: Photographic Evidence and the Power of Analogy' (1998) 10 (1) *Yale Journal of Law and the Humanities* 1.

5 Henry A. Field Jr., 'Uses and Limitations of X-Ray Pictures as Evidence' (1967) 2(4) *Forum* 219.

6 'First Radiograph in Evidence' (1898) 2 *American X-Ray Journal* 155.

technology to make 'it possible to look beneath the tissues of the human body' and reveal the 'hidden mysteries' of organs, tissue and bone.[7]

The first two decades following the discovery of X-rays were unnerving for legal institutions. Early adopters were mostly photographers, who conceptualised 'roentgenography' as a sub-discipline of photography and set up 'X-ray studios' as a form of entertainment for mainly urban dwellers.[8] Yet the comparison between X-rays and photography was viewed as misleading, because the former did not so much produce photographs of an object, prominent lawyers argued, as they revealed 'shadowgraphs'. The issue here was whether an X-ray image replicated the interiority of an object, which could not be verified by the human eye, or represented it through a play of shadows:

> It follows from this that the use of the erroneous term 'X-ray photograph,' therefore, misleads jurors and others to regard the appearance on an X-ray film as comparable, in many respects, to a photograph. ... [It is rather] nothing more or less than a picture of shadows, and if the term shadow-graph or skiagraph were used, it would give a much clearer idea and would perhaps afford a better understanding of its legal value.[9]

The future of the X-ray image as an evidentiary tool thus seemed uncertain in the first two decades of the twentieth century. 'There was no readily applicable body of medical knowledge', Bernike Pasveer writes, 'on the basis of which to assess either normal or pathological appearances; there were no implicit meanings in the pictures'.[10] If the skiagraph did not provide direct evidence of the inner cavities of an object, but rather depicted its interiority through the aesthetics of chiaroscuro, before the image could be accepted as evidence in a court of law, its meaning had to be made sense of by an expert trained in the hermeneutics of roentgenography. The call made by judicial officers for medical experts to verify X-ray images – that is, to decipher the meanings of shadows and explain the way that they became imprinted on film – paralleled the development of radiology as a specialised discipline of medicine in the 1910s and 1920s. Only a medical practitioner trained as a radiologist could competently produce, interpret and verify the meaning of X-ray images.

7 'Smith v Grant' (1896) 29 *Chicago Legal News* 145 quoted in S.W. Donaldson, 'Roentgeno-grams as Evidence' (1938) 1 *American Journal of Medical Jurisprudence* 228, 228. The case was held in the District Court of Denver, Colorado.

8 Tal Golan, *Laws of Men and Laws of Nature: The History of Scientific Expert Testimony in England and America* (Harvard University Press, 2004) 190. See also José van Dijck, *The Transparent Body: A Cultural Analysis of Medical Imaging* (University of Washington Press, 2005) 92.

9 Orlando F. Scott, 'Rontgenograms and their Chrono-logic Legal Recognition' (1929) 24 *Illinois Law Review* 674, 674. See also R. Harvey Reed, 'The X-Ray from a Medico-legal Standpoint' (1898) 30(18) *Journal of the American Medical Association* 1013; L.H. Garland, 'The Interpretation of X-rays in Court Hearings' (1938) 1 *American Journal of Medical Jurisprudence* 19.

10 Bernike Pasveer, 'Knowledge of Shadows: The Introduction of X-Ray Images in Medicine' (1989) 11(4) *Sociology of Health and Illness* 360, 363.

Medico-legal professionals quickly saw the potential role that radiography could play in the death investigation process. It was touted as useful for determining the identity of a corpse, the age of the deceased, the location of bone fractures and even that death had occurred – '[a]fter death, the pulsation of the heart is invisible, and the organ presents a sharp outline'.[11] While, at first, X-rays could only project shadows of 'the skeleton, large dense organs such as the lungs, and heavy foreign objects such as bullets', with technological advancements, they soon came to represent 'the respiratory system, the circulatory system, the alimentary tract and many organs of the body'.[12] Radiographers praised shadowgraphs as more accurate than the oral testimony of the human witness. Shadowgraphs could pinpoint bullets and other foreign objects lodged in bones, organs or muscles, without compromising the integrity of the evidence. They could show the 'character and extent of an injury', such as the '[r]upture of ligaments and cartilages, tendons, diseased conditions of the bones', prior to any dissection of the corpse.[13]

The most common use of forensic radiography was initially post-mortem identification, especially for drowned, charred or disfigured remains.[14] This practice evolved from the science of retrospectroscopy. Retrospectroscopy required the comparison of two or more X-ray images of the same patient to improve the diagnostic process and detect pathological morbidities that remained invisible to the human eye:

> The Röntgenologist made an X-ray picture of a patient, but refrained from making a definitive diagnosis on the basis of it. If the patient under examination died or had to be operated on, the X-ray worker attended the operation or autopsy, and in the latter case he sometimes made Röntgen photographs of the dead organs. Thus he became acquainted with what the 'true' diagnosis should have been.[15]

The development of *forensic* retrospectroscopy, however, involved comparing post-mortem shadowgraphs not only with X-ray images taken during the

11 Mihran K. Kassabian, 'The Roentgen Rays in Forensic Medicine' (1901) 19 *Medico-legal Journal* 407, 417.
12 Charles C. Scott, 'X-Ray Pictures as Evidence' (1946) 44(5) *Michigan Law Review* 773, 773. In the early twentieth century, X-ray images were often clouded by distortions, deformations and magnification caused by the incipient technology. This was remedied partly by the use of 'contrast media', which created differing densities and heightened shadows, for organs, cavities or tracts affected by the use of such substances.
13 Kassabian, above n 11, 408 and 414.
14
> Gustav Bucky presented before the 1913 Roentgen congress a detailed reconstruction of a crime from examination of a corpse that had been buried for four years. ... Bucky also showed the usefulness of x-ray examination of the ash residue of burned corpses.
> Vincent D. Collins, 'Origins of Medico-Legal and Forensic Roentgenology', in A.J. Bruwer (ed), *Classic Descriptions in Diagnostic Roentgenology* (C.C. Thomas, 1964) Vol. 2, 1601

15 Pasveer, above n 10, 370.

deceased's lifetime, but also with dental records, clinical files and medical photographs. In other words, identifying a corpse by using radiography was only made possible by the collection, storage and retrieval of a wide range of ante-mortem records. As Harry Levine, III explains, the dentist assumed a critical role in curating this archive, which could ideally be accessed by medico-legal professionals when seeking to determine the identity of a corpse:

> The enamel on teeth absorbs even more X-rays than bone and cavities and malformations were obvious on the plates. Personal dental history was unique to the individual. Thus, dentists accumulated the first archives of medical imaging records that could identify individuals even after death. Dental records and X-rays are important tools for modern coroners. The first recorded use of X-rays to identify the dead was in the aftermath of a fire at a benefit bazaar in Paris, France, on 4 May 1897, in which more than two hundred socially prominent patrons died. X-rays were used to identify the bodies that were charred beyond recognition.[16]

Identification of a corpse was also possible by comparing the X-rays of sinus and mastoid bones of the deceased with their ante-mortem medical records, as in the following case:

> Fingerprints on record would have been of no aid, for one arm was gone and there was scarcely any flesh on the bones of the other hand. The facial bones were bare of flesh. However, an examination of the sinuses and mastoids with a portable [X-ray] machine at the morgue and comparisons of the resulting films with the plates on file, established beyond any possibility of doubt that both sets of roentgenograms were of one and the same individual. The operation on the left mastoid following the first examination had destroyed its value for identification, but as both sides had been examined, the right mastoid was available for comparison. Thirteen points of identity in the sinuses and seven in the right mastoid were noted; the number could have been extended indefinitely.[17]

---

16 Harry Levine, III, *Medical Imaging* (ABC-CLIO, Santa Barbara, 2010) 110. See also William G. Eckert and Neil Garland, 'The History of the Forensic Applications in Radiology' (1984) 5(1) *The American Journal of Forensic Medicine and Pathology* 53.

17 Frederick M. Law, 'Roentgenograms as a Means of Identification' (1934) 26 *American Journal of Surgery* 195 quoted in Charles C. Scott, 'X-Ray Pictures as Evidence' (1946) 44(5) *Michigan Law Review* 773, 786 and 787. Scott also discusses the development of 'dactyloscopic radiography' in the mid-twentieth century, which involved imaging the 'fingerprints of a corpse that is in an advanced state of putrefaction':

> If sufficiently advanced, decomposition will render the skin tissue so macerated that any attempt to make an ink impression of the fingers will yield blurred results or even cause the skin to peel off and adhere to the inking plate. Under these conditions a properly taken radiograph of the hand will reproduce the ridge pattern so perfectly that it can be enlarged and compared with other fingerprints to establish identity.

Forensic radiography was undoubtedly considered most valuable for determining the identity of an unknown corpse. But it was also quickly seen as beneficial for distinguishing between ante-mortem and post-mortem damage. In the nineteenth century, the only way to differentiate injuries inflicted before death from those caused by decomposition was 'to cut deep into the affected flesh in order to observe directly underlying tissue and the condition of the blood'.[18] This changed, of course, when pathologists were able to compare pre- and post-death shadowgraphs to discern the existence of any ante-mortem bruises, fractures or wounds prior to opening up the body.[19] The invention of radiography in 1895 thus enabled the development of a number of optical apparatuses that would come to redefine the conduct of the post-mortem examination in the twentieth century.

The consequences of this paradigmatic shift in a visual hermeneutics of the corpse are not to be underestimated. Human dissection was central to the transmogrifications of medical epistemology in the nineteenth century. The isolation of disease on 'tissular surfaces', as Foucault reminds us, was only made possible by gazing into the inner crevices of the corpse. Only by cutting deep into putrefying flesh could the physician discern a truth yielded by the morbid appearance of organs. The medical gaze reduced organic densities to tissular surfaces, which could then be made sense of as pathological in themselves. But this gaze changed once again in the twentieth century due to the emergence of new technologies for screening the corpse. Radiological vision offered the physician a different optical apparatus for revealing a truth of morbidity. The human body could now be classified and measured by the penetration of radiating light and the play of chiaroscuro on photographic film.

Foucault did not discuss the significance of the late nineteenth-century discovery of X-rays in *The Birth of the Clinic*. However, he did identify the stethoscope as a 'semi-tactile, semi-auditory' apparatus that 'solidified distance' between the physician and the human body and, in a comparable way to the X-ray, altered the medical gaze by making 'it possible to fix the virtual image of what is occurring well below the visible area'.[20] The similarities between the stethoscope and the X-ray are not obvious. The former augmented an auditory-tactile relation with the patient, while the latter emphasised the primacy of vision in the diagnostic process. Nonetheless, both apparatuses, alongside other medical devices invented in the late nineteenth century, such as the ophthalmoscope and laryngoscope, enabled 'practitioners to extend the power of their senses into the

---

18 Ian Burney and Neil Pemberton, 'Bruised Witness: Bernard Spilsbury and the Performance of Early Twentieth-Century English Forensic Pathology' (2011) 55 *Medical History* 41, 48.

19 X-rays promised to visualise beyond the damage to the corpse wrought by putrefaction, which 'generate[d] new signs that made post-mortem changes indistinguishable from pre-mortem ones, while at the same time erasing evidentially significant pre-mortem marks-bruises ... ': Ibid, 49.

20 Michel Foucault, *The Birth of the Clinic: An Archaeology of Medical Perception* (Alan Sheridan trans, Routledge Classics, 2003) 202.

body's inner cavities'.[21] These devices all revealed in different ways, using different senses, but at a similar *distance* from the body, what lies hidden underneath its surfaces.

The X-ray altered the coroner's view of the corpse in the twentieth century. As I discussed in an earlier chapter, the performance of *super visum corporis* – the legal duty to view the corpse – was necessary in the nineteenth century for an inquest to proceed. By the mid-twentieth century, the coronial jury ceased to view the dead body during an inquest, while the post-mortem examination was left almost entirely in the hands of the forensic pathologist. While coroners could still, if they desired to do so, gaze upon the surface of the corpse as it lay in the morgue, their relations with the dead were increasingly mediated by a medical expert. The X-ray augmented this distancing between the coroner and the corpse. Its connotations for the death investigation process could only be interpreted by a trained specialist, who was proficient in its visual hermeneutics. Forensic radiographers thus assumed the role of mediator between the mechanical production of shadow images and the curation of an institutional narrative of death causation. They became translators of 'the visual into the verbal ... the corporeal into the linguistic, in the textual genre of the autopsy protocol or report, that finally establishes the evidentiary value of the body of the victim'.[22]

This is not all that can be said about the discursive effects of radiography on the coroner's view of the corpse. Or, to put this differently, the forensic radiographer was not the only force, and perhaps not even the most important one, that intervened in the ocular relations between the coroner and the dead body in the twentieth century. The X-ray in itself 'mechanised' the forensic gaze.[23] The coroner's view became mediated by a screening technology that differed substantially from the pathologist's scalpel. This technology relied less on the fallibility of the medical expert, and more on the 'mechanical objectivity' of a machine-based process: 'Of all the nineteenth-century inventions made available for forensics', Piyel Haldar writes, 'the x-ray was the one device that attempted to completely replace the fallible human witness to an extent ... with a process'.[24] The mechanisation of the forensic gaze was undoubtedly intended to make images of the corpse more 'objective'. But, in doing so, it produced 'an image that could easily belong to any*body*'.[25] In other words, it transformed the particular into the general, and created an ideal image of the interiority of a unique body, which could be used by coroners, pathologists and radiographers for the purposes of comparison, classification and standardisation. Medical imaging

21 Golan, above n 8, 180.
22 Joseph Pugliese, '"*Super Visum Corporis*": Visuality, Race, Narrativity and the Body of Forensic Pathology' (2002) 14(2) *Law and Literature* 367, 369, 384.
23 The word 'mechanised' is borrowed from José van Dijck's insightful analysis of a cultural history of medical imaging technology: above n 8, 99.
24 Piyel Haldar, 'Forensic Representations of Identity: The *Imago*, the X-Ray and the Evidential Image' (2013) 7(2) *Law and Humanities* 129, 141 (emphasis in original).
25 Ibid, 146 (emphasis in original).

technology paradoxically provided the worlds of law and medicine with tools for identifying a corpse, by its teeth, sinus and mastoid bones, fingerprints and bone fractures, while also transforming its uniqueness into generalities that could be applied for other purposes.

It is possible to conceive that both the forensic radiographer and X-ray technology ossified a distance in the death investigation process between the coroner and the corpse, a remoteness that is believed by medical historians to have been in train since the medicalisation of the inquest in the mid-nineteenth century. However, this theory relies on the assumption that the X-ray was successful in supplanting 'subjective sensorial impressions by supposedly objective visual evidence'.[26] And it ignores the idea that the radiographer's interpretation of an image of shadows, just like the pathologist's and coroner's, was dependent on the intersubjective observations of a machine-based process:

> The body – made transparent by a host of new mechanical instruments – is anything but an objective object of study. On the contrary, X-ray pictures, like other mechanical reproductions, always yield mediated perspectives, as their meanings are always shaped by the knowledge and feelings of their interpreters. The X-ray, as the 'enhanced eye' of art and science, is a small but significant element in the emergence of a new way of seeing, a visual regime that believes in transparency and translucency.[27]

The X-ray, therefore, did not grant forensic pathologists exclusive control over interpretations of images of the corpse.[28] Even though the medical expert played an important role in translating the meanings of shadowgraphs before an inquisitorial forum, coroners still had to develop new aptitudes for viewing the interiority of the corpse from afar and new skills for interpreting the legal significance of shadows as evidence of death causation. For only by acquiring expertise in understanding the transformations of a visual hermeneutics of the corpse could coroners attach its image to the institutional life of coronial law.

What this history of radiography reveals then, is that it is important, when new technologies are introduced in the death investigation process, to examine their effects on how office-holders take care of the dead. In other words, the question here is not whether medical imaging technologies marginalise the role that coroners assume in the death investigation process, but rather how they affect the *quality* of caring practices. This quality may vary whether one is

---

26 van Dijck, above n 8, 86.
27 Ibid, 99.
28 Rebecca Scott Bray argues that forensic pathologists produce 'a number of medico-legal portraits of the dead' and 'despite the seeming stability of these images, this "reality" can be challenged by different interpretations of injury, thereby undoing stable meaning and potentially rewriting death': Rebecca Scott Bray, 'Fugitive Performances of Death and Injury' (2006) 10 *Law Text Culture* 41, 43. See also Rebecca Scott Bray, *The Eschatology of the Image* (PhD Thesis, University of Melbourne, 2001) 128–129.

cutting up the body, peering into open wounds or deciphering shadows on a translucent image. The quality of care depends on how coroners are attuned to the way technologies mediate legal relations with the dead, how technologies attach the dead to institutional formations of law and how these technologies demand from office-holders the development of an ethical mindset. By the mid-twentieth century, the mechanised gaze was intractably attached to the institutional life of coronial law. Its use during the post-mortem examination became almost compulsory, such that every mortuary required a portable X-ray machine.[29] The X-ray became a key example of how caring practices changed from the late nineteenth century, and even in some cases supplemented the autopsy, for detecting what could not be seen by the human eye: a history of bone fractures; the location of minuscule foreign objects, which required careful removal; and particular causes of death, such as strangulation, subarachnoid haemorrhage and air embolism.

Coda

Taking care of the dead requires an awareness of how the dead persist in a legal institution. It is only by recognising the persistence of what remains that an institution can learn how to cultivate relations of care with the dead. The quality of caring practices, however, has been shown in this chapter to differ according to the technologies that mediate relations between the living and the dead. Even if a corpse is replaced by its image, and that image is abstracted from what makes it particular, the dead still demand that the living assume responsibility for their 'afterlife'. Afterlife is neither the cessation of life nor the spiritual presence of the dead. It is, as Robert Pogue Harrison writes, 'the place of that persistence [that] remains open'.[30]

Law and the Dead: Technology, Relations and Institutions has narrated a history of the institutional life of coronial law. It has offered readers a historical account of how coroners occupied their offices in the nineteenth and twentieth centuries. The book has traced how jurisdictional technologies attached the dead to the conduct of coronial law and its institutional formations. Each chapter describes how a different technology shaped the way coroners formed legal relations with the dead in the nineteenth and twentieth centuries. Coroners assumed a responsibility of care, by making a place for the dead in their lives, speaking on behalf of those who could not speak, recording biographies of the deceased and discerning in shadows a narrative of death causation. The question that unites each chapter is: how did coroners engage with the dead who were brought before their laws? In other words, all chapters have been guided by the

29 Bernard Knight, *The Post-Mortem Technician's Handbook: A Manual of Mortuary Practice* (Blackwell Scientific Publications, Oxford, 1984) 202.
30 Robert Pogue Harrison, *The Dominion of the Dead* (The University of Chicago Press, 2003) 154.

problem of how technology mediated the forms by which coroners encountered the dead.

This book has explored an array of technologies to show what sustained the institutional life of coronial law in the nineteenth and twentieth centuries. The technologies analysed herein are far from exhaustive. They were chosen not only because they are historically significant for understanding how the office of coroner was institutionalised over the past two centuries, but also for comprehending how this office has come to assume the important role that it now occupies in contemporary society. It is hoped that this book tempts readers to think differently about how a legal institution, which underwent significant change in the nineteenth and twentieth centuries, implemented new technologies for taking care of the dead. The book has shown how coroners cultivated an ethics of responsibility for living with the dead. It is not my intention, though, to put forward this office as a paragon of how a legal institution ought to form legal relations with the dead. The history of the office of coroner has made this abundantly clear. But I do hope that we can learn from this complicated history what is at stake when a legal institution learns to live with the dead.

Bibliography

Agamben, Giorgio, *Language and Death: The Place of Negativity* (Karen E. Pinkus and Michael Hardt trans, University of Minnesota Press, 1991).

Agamben, Giorgio, *Nudities* (David Kishik and Stefan Dedatella trans, Stanford University Press, 2011).

Agamben, Giorgio, *Opus Dei: An Archaeology of Duty* (Adam Kotsko trans, Stanford University Press, 2013).

Anderson, Olive, *Suicide in Victorian and Edwardian England* (Clarendon Press, 1987).

Anon., 'First Radiograph in Evidence' (1898) 2 *American X-Ray Journal* 155.

Ariès, Philippe, *Western Attitudes toward Death: From the Middle Ages to the Present* (Patricia M. Ranum trans, The Johns Hopkins University Press, 1974).

Ariès, Philippe, *The Hour of Our Death: The Classic History of Western Attitudes toward Death over the Last One Thousand Years* (Helen Weaver trans, Vintage Books, 2nd edn, 2008).

Arnold, Catharine, *Necropolis: London and Its Dead* (Pocket Books, 2007).

Asen, Daniel, 'Song Ci (1186–1249), "Father of World Legal Medicine": History, Science, and Forensic Culture in Contemporary China' (2017) 11 *East Asian Science, Technology and Society: An International Journal* 185.

Baker, William, *A Practical Compendium of the Recent Statutes, Cases and Decisions Affecting the Office of the Coroner* (Butterworths, 1851).

Barr, Olivia, *A Jurisprudence of Movement: Common Law, Walking, Unsettling Place* (Routledge, 2016).

Barry, Andrew, *Political Machines: Governing a Technological Society* (The Athlone Press, 2001).

Bataille, Georges, *The Accursed Share: An Essay on General Economy – Volume 1: Consumption* (Robert Hurley trans, Zone Books, 1991).

Bataille, Georges, *Theory of Religion* (Robert Hurley trans, Zone Books, 1992).

Baudrillard, Jean, *Symbolic Exchange and Death* (Iain Hamilton Grant trans, Sage Publications, 1993).

Bauman, Zygmunt, *Mortality, Immortality and Other Life Strategies* (Polity Press, 1992).

Bayatrizi, Zohren, *Life Sentences: The Modern Ordering of Mortality* (University of Toronto Press, 2008).

Becker, Ernest, *The Denial of Death* (Free Press Paperbacks, 1973).

Benjamin, Walter, *Illuminations: Essays and Reflections* (Harry Zohn trans, Schocken Books, 1968).

Bennett, J.M., 'The Evolution of Court Houses in New South Wales', in Terry Naughton (ed), *Places of Judgment: New South Wales* (Law Book Co., 1987).

Bennett, Richard E., *Capital Punishment and the Criminal Corpse in Scotland, 1740–1834* (Palgrave Macmillan, 2018).

Biber, Katherine and Trish Luker, 'Evidence and the Archive: Ethics, Aesthetics, and Emotion' (2014) 40(1) *Australian Feminist Law Journal* 1.

Blanco, María del Pilar and Esther Peeren (eds), *The Spectralities Reader: Ghosts and Hauntings in Contemporary Cultural Theory* (Bloomsbury, 2013).

Boys, William F.A., *A Practical Treatise on the Office and Duties of Coroners in Upper Canada: With an Appendix of Forms* (W.C. Chewett & Co., 1864).

Brazil, Raymond, 'Respecting the Dead, Protecting the Living' (2008) 12(SE2) *Australian Indigenous Law Review* 45.

Bronfen, Elisabeth, *Over Her Dead Body: Death, Femininity and the Aesthetic* (Manchester University Press, 1992).

Brown, James Drysdale, *A Short Manual: For the Guidance of Coroners, Deputy Coroners, and Justices Acting as Coroners, in Victoria* (Government Printer, 1911).

Brown-May, Andrew, 'History and Development of the Site', in Andrew Brown-May and Norman Day (eds), *Federation Square* (Hardie Grant Books, 2003).

Brown-May, Andrew and Simon Cooke, 'Death, Decency and the Dead-House: The City Morgue in Colonial Melbourne' (2004) 3 *Provenance: the Journal of Public Record Office Victoria* 4. http://prov.vic.gov.au/publications/provenance/proven ance2004/death-decency-and-the-dead-house

Burn, Richard J., *Justice of the Peace and Parish Officer, Volume II* (Sweet, Maxwell and Son, and Steven's and Norton, [1755] 1845 edn).

Burney, Ian, 'Making Room at the Public Bar: Coroners' Inquests, Medical Knowledge and the Politics of the Constitution in Early-Nineteenth-Century England', in James Vernon (ed), *Re-Reading the Constitution: New Narratives in the Political History of England's Long Nineteenth Century* (Cambridge University Press, 1996).

Burney, Ian, *Bodies of Evidence: Medicine and the Politics of the English Inquest, 1830–1926* (Johns Hopkins University Press, 2000).

Burney, Ian, *Poison, Detection and the Victorian Imagination* (Manchester University Press, 2012).

Burney, Ian A., 'Viewing Bodies: Medicine, Public Order, and English Inquest Practice' (1994) 2(1) *Configurations* 33.

Burney, Ian A., 'A Poisoning of No Substance: The Trials of Medico-Legal Proof in Mid-Victorian England' (1999) 38(1) *Journal of British Studies* 59.

Burney, Ian A. and Neil Pemberton, 'Bruised Witness: Bernard Spilsbury and the Performance of Early Twentieth-Century English Forensic Pathology' (2011) 55 *Medical History* 41.

Burton, Antoinette (ed), *Archive Stories: Facts, Fiction and the Writing of History* (Duke University Press, 2005).

Butler, Judith, '"What Is Critique?" an Essay on Foucault's Virtue', in David Ingram (ed), *The Political Blackwell Readings in Continental Philosophy* (Wiley-Blackwell, 2002).

Butler, Sara M., *Forensic Medicine and Death Investigation in Medieval England* (Routledge, 2015).

Caffyn, Mannington, 'They Met at the Morgue', *The Bulletin* (Melbourne), 9 July 1892, 21.

Campbell, Enid, 'Inferior and Superior Courts and Courts of Record' (1997) 6 *Journal of Judicial Administration* 249.

Cantor, Norman L., *After We Die: The Life and Times of the Human Cadaver* (Georgetown University Press, 2010).

Carpenter, Belinda and Gordon Tait, 'The Autopsy Imperative: Medicine, Law and the Coronial Investigation' (2010) 31 *Journal of Medical Humanities* 205.

Carter, Paul, *Mythform: The Making of Nearamnew at Federation Square* (Miegunyah Press, 2005).

Carter, Paul, *The Road to Botany Bay: An Exploration of Landscape and History* (University of Minnesota Press, 2010).

Cawthorn, Elizabeth, 'New Life for the Deodand: Coroners' Inquests and Occupational Deaths in England, 1830–46' (1989) 33(2) *American Journal of Legal History* 137.

Cawthorn, Elizabeth, *Job Accidents and the Law in England's Early Railway Age* (Edwin Mellen Press, 1997).

Challinger, Michael, *Historic Court Houses of Victoria* (Palisade Press, 2001).

Choron, Jacques, *Death and Western Thought* (Collier Books, 1963).

Clanchy, M.T., *From Memory to Written Record: England 1066–1307* (Blackwell, 2nd edn, 1993).

Clark, David (ed), *The Sociology of Death: Theory, Culture and Practice* (Blackwell, 1993).

Clark, Michael and Catherine Crawford (eds), *Legal Medicine in History* (Cambridge University Press, 1994).

Collins, Vincent D., 'Origins of Medico-Legal and Forensic Roentgenology', in A. J. Bruwer (ed), *Classic Descriptions in Diagnostic Roentgenology* (C.C. Thomas, 1964) Vol. 2.

Condren, Conal, *Argument and Authority in Early Modern Europe: The Presupposition of Oaths and Offices* (Cambridge University Press, 2006).

Conway, Heather, *The Law and the Dead* (Routledge, 2016).

Cooke, Simon, *Secret Sorrows: A Social History of Suicide in Victoria, 1824–1921* (PhD Thesis, University of Melbourne, 1998).

Cooke, Simon, *Inquests* (2008) The Encyclopedia of Melbourne Online. www.emelbourne.net.au/biogs/EM00756b.htm

Corbin, Alain, *The Foul and the Fragrant: Odor and the French Social Imagination* (Miriam L. Kochan trans, Harvard University Press, 1986).

Cordner, Stephen and Fiona Leahy, 'Forensic Medicine and the Supreme Court', in Simon Smith (ed), *Judging for the People: A Social History of the Supreme Court in Victoria 1841–2016* (Allen & Unwin, 2016).

The Coroner's Guide: Or, the Office and Duty of a Coroner: Containing Variety of Precedents, and Proper Instructions for Executing the Said Office. Compiled from the Best Authorities (John Worrall, 1756).

Crawford, Catherine, 'Medicine and the Law', in W.F. Bynum and Roy Porter (eds), *The Companion Encyclopaedia of the History of Medicine* (Routledge, 1993).

Crawford, Catherine, 'Legalizing Medicine: Early Modern Legal Systems and the Growth of Medico-Legal Knowledge', in Michael Clark and Catherine Crawford (eds), *Legal Medicine in History* (Cambridge University Press, 1994).

Crowley, Patrick R., 'Roman Death Masks and the Metaphorics of the Negative' (2016) 64 *Grey Room* 64.

Daston, Lorraine and Peter Galison, *Objectivity* (Zone Books, 2010).

de Certeau, Michel, *The Practice of Everyday Life* (Steven Rendall trans, University of California Press, 1984).

'Debate on Coroners in the Legislative Assembly Tuesday 2nd October 1877' 1877 22(10) *Australian Medical Journal* 305.

Derrida, Jacques, *Aporias* (Thomas Dutoit trans, Stanford University Press, 1993).

Derrida, Jacques, *The Gift of Death* (David Wills trans, The University of Chicago Press, 1995).

Derrida, Jacques, *Specters of Marx: The State of the Debt, the Work of Mourning and the New International* (Peggy Kamuf trans, Routledge Classics, 2006).

Derrida, Jacques et al., *Ghostly Demarcations: A Symposium on Jacques Derrida's Specters of Marx* (Verso, [1999] 2008 edn).

Dollimore, Jonathan, *Death, Desire and Loss in Western Culture* (Routledge, 1998).

Donaldson, S.W., 'Roentgenograms as Evidence' (1938) 1 *American Journal of Medical Jurisprudence* 228.

Dorsett, Shaunnagh and Ian Hunter (eds), *Law and Politics in British Colonial Thought: Transpositions of Empire* (Palgrave Macmillan, 2010).

Dorsett, Shaunnagh and John McLaren (eds), *Legal Histories of the British Empire: Laws, Engagements, Legacies* (Routledge, 2014).

Dorsett, Shaunnagh and Shaun McVeigh, 'The *Persona* of the Jurist in Salmond's *Jurisprudence*: On the Exposition of "What Law Is ..."' (2007) 38 *Victoria University of Wellington Law Review* 771.

Dorsett, Shaunnagh and Shaun McVeigh, *Jurisdiction* (Routledge, 2012).

Drakopoulou, Maria, 'Of the Founding of Law's Jurisdiction and the Politics of Sexual Difference: The Case of Roman Law', in Shaun McVeigh (ed), *Jurisprudence of Jurisdiction* (Routledge, 2007).

du Gay, Paul, *In Praise of Bureaucracy: Weber, Organization, Ethics* (Sage Publications, 2000).

du Gay, Paul, *Organizing Identity* (Sage Publications, 2007).

du Gay, Paul, 'Max Weber and the Moral Economy of Office' (2008) 1(2) *Journal of Cultural Economy* 129.

Duncanson, Ian, *Historiography, Empire and the Rule of Law: Imagined Constitutions, Remembered Legalities* (Routledge, 2012).

Durkheim, Émile, *Suicide: A Study in Sociology* (John A. Spaulding and George Simpson trans, Routledge, 2005).

Eckert, William G. and Neil Garland, 'The History of the Forensic Applications in Radiology' (1984) 5(1) *The American Journal of Forensic Medicine and Pathology* 53.

Elmslie, Ronald and Susan Nance, 'Smith, William Ramsay (1859–1937)', in *Australian Dictionary of Biography* (Melbourne University Press, 1988) Vol. 2. http://adb.anu.edu.au/biography/smith-william-ramsay-849

Engels, Frederick, *The Condition of the Working-Class in England in 1844* (Florence Kelley Wischnewetzky trans, George Allen & Unwin, 1892).

Esposito, Roberto, 'The *Dispositif* of the Person' (2012) 8(1) *Law, Culture and the Humanities* 17.

Esposito, Roberto, *Persons and Things: From the Body's Point of View* (Zakiya Hanafi trans, Polity Press, 2015).

Farge, Arlene, *The Allure of the Archives* (Thomas Scott-Railton trans, Yale University Press, 2015).

Farr, Samuel, 'Elements of Medical Jurisprudence', in Thomas Cooper (ed), *Tracts on Medical Jurisprudence* (James Webster, [1788] 1819 edn).

Featherstone, Mike, 'Archive' (2006) 23(2–3) *Theory, Culture & Society* 591.

Feifel, Herman, *The Meaning of Death* (McGraw-Hill, 1st edn, 1959).

Ferguson, Kathryn, 'Imagining Early Melbourne' (2004) 1(1) *Postcolonial Text* 1.

Fforde, Cressida, 'From Edinburgh University to the Ngarrindjeri Nation, South Australia' (2009) 61(1–2) *Museum International* 41.

Fforde, Cressida, Jane Hubert and Paul Turnbull (eds), *The Dead and Their Possessions: Repatriation in Principle, Policy and Practice* (Routledge, 2002).

Field, Jr., Henry A., 'Uses and Limitations of X-Ray Pictures as Evidence' (1967) 2(4) *Forum* 219.

Fink, William, 'The O'Callaghan Baby Case [From the "Australian Law Times"]', *Table Talk* (Melbourne), 25 August 1899, 3.

Finkelstein, Jacob, 'The Goring Ox: Some Historical Perspectives on Deodands, Forfeitures, Wrongful Death and the Western Notion of Sovereignty' (1972–1973) 46(2) *Temple Law Quarterly* 169.

Finn, Paul, 'The Law and Officials', in R.A. Chapman (ed), *Ethics in Public Service* (Edinburgh University Press, 1993).

Finnane, Mark and Jonathan Richards, '"You'll Get Nothing Out of It"? the Inquest, Police and Aboriginal Deaths in Colonial Queensland' (2004) 123(35) *Australian Historical Studies* 84.

Fitzherbert, Sir Anthony, *In This Boke Is Conteyned Ye Office of Shyryffes, Baylyffes of Lybertyes, Escheatours, Constables, & Coroners: And Sheweth What Euerye One of Them May Do by Vertue of Theyr Offyces Drawen Out of Bokes of the Comen Lawe & of the Statutes* (Wyllyam Powell, 1549).

Forbes, Thomas Rogers, 'Crowner's Quest' (1978) 68(1) *Transactions of the American Philosophical Society*.

Ford, Richard T., 'Law's Territory: A History of Jurisdiction' (1999) 97(4) *Michigan Law Review* 843.

Foucault, Michel, *The Archaeology of Knowledge & the Discourse on Language* (A.M. Sheridan Smith trans, Pantheon Books, 1972).

Foucault, Michel, 'Nietzsche, Genealogy, History', in David F. Bouchard (ed), *Language, Counter-Memory, Practice: Selected Essays and Interviews* (Cornell University Press, 1977).

Foucault, Michel, 'The Confession of the Flesh', in Colin Gordon (ed), *Power/ Knowledge: Selected Interviews and Other Writings, 1972–1977* (Colin Gordon trans, Harvester Press, 1980).

Foucault, Michel, 'Of Other Spaces' (Jay Miskowiec trans, 1986) 16 (1) *Diacritics* 22.

Foucault, Michel, 'The Political Technology of Individuals', in Luther H. Martin, Huck Gutman and Patrick H. Hutton (eds), *Technologies of the Self: A Seminar with Michel Foucault* (Tavistock, 1988).

Foucault, Michel, *Discipline and Punish: The Birth of the Prison* (Alan Sheridan trans, Penguin Books, 1991).

Foucault, Michel, *The History of Sexuality, Volume 1: The Will to Knowledge* (Robert Hurley trans, Penguin Books, 1998).

Foucault, Michel, 'The Lives of Infamous Men', in James D. Faubion (ed), *Power: Essential Works of Foucault 1954–1984, Volume 3* (Robert Hurley trans, Penguin Books, 2002).

Foucault, Michel, *Abnormal: Lectures at the Collège De France 1974–1975* (Graham Burchell trans, Picador, 2003).

Foucault, Michel, *The Birth of the Clinic: An Archaeology of Medical Perception* (Alan Sheridan trans, Routledge Classics, 2003).

Foucault, Michel, *Death and the Labyrinth: The World of Raymond Roussel* (Charles Ruas trans, Continuum, 2004).

Foucault, Michel, *Speech Begins after Death: In Conversation with Claude Bonnefoy* (Robert Bononno trans, University of Minnesota Press, 2013).

Freckelton, Ian and David Ranson, *Death Investigation and the Coroner's Inquest* (Oxford University Press, 2006).

Freud, Sigmund, *Beyond the Pleasure Principle, Group Psychology and Other Works, Volume XVII (1920–1922)* (James Strachey trans, Vintage, The Hogarth Press and the Institute of Psycho-Analysis, 2001).

Freud, Sigmund, *On the History of the Psycho-Analytic Movement, Papers on Metapsychology and Other Works, Volume XIV (1914–1916)* (James Strachey trans, Vintage, The Hogarth Press and the Institute of Psycho-Analysis, [1915] 2001 edn).

Frost, Michael H., *Introduction to Classical Legal Rhetoric: A Lost Heritage* (Routledge, 2005).

Fuss, Diana, 'Corpse Poem' (2003) 30(1) *Critical Inquiry* 1.

Garland, L.H., "The Interpretation of X-Rays in Court Hearings" (1938) 1 *American Journal of Medical Jurisprudence* 19.

Geary, Patrick J., *Living with the Dead in the Middle Ages* (Cornell University Press, 1994).

Gitelman, Lisa, *Paper Knowledge: Toward a Media History of Documents* (Duke University Press, 2014).

Gittings, Clare, *Death, Burial and the Individual in Early Modern England* (Routledge, 1984).

Golan, Tal, *Laws of Men and Laws of Nature: The History of Scientific Expert Testimony in England and America* (Harvard University Press, 2004).

Golder, Hilary, *High and Responsible Office: A History of the NSW Magistracy* (Oxford University Press, 1991).

Goodrich, Peter, 'Rhetoric as Jurisprudence: An Introduction to the Politics of Legal Language' (1984) 4 *Oxford Journal of Legal Studies* 84.

Goodrich, Peter, *Reading the Law: A Critical Introduction to Legal Method and Techniques* (Basil Blackwell, 1986).

Goodrich, Peter, 'Specula Laws: Image, Aesthetic and Common Law' (1991) 2(2) *Law and Critique* 233.

Goodrich, Peter, 'Visive Powers: Colours, Trees and Genres of Jurisdiction' (2008) 2 (2) *Law and Humanities* 213.

Goodrich, Peter, 'The Theatre of Emblems: On the Optical Apparatus and the Investiture of Persons' (2012) 8(1) *Law, Culture and the Humanities* 47.

Gordon, Bruce and Peter Marshall (eds), *The Place of the Dead: Death and Remembrance in Late Medieval and Early Modern Europe* (Cambridge University Press, 2000).

Gorer, Geoffrey, *Death, Grief and Mourning in Contemporary Britain* (Cresset Press, 1965).

Graham, Clare, 'Sudden Death and the LCC: Accommodation for Inquests in London before the First World War' (1995) 1 *Arq* 60.

Graham, Clare, *Ordering Law: The Architectural and Social History of the English Law Court* (Ashgate, 2003).

Gray, Adrian, 'A Review of Transport and the Law of Deodand' (2011) 212 *Journal of the Railway and Canal Historical Society* 26.

Gross, Charles (ed), *Select Cases from the Coroner's Rolls, A.D. 1265–1413* (Bernard Quaritch, 1896).

Halbwachs, Maurice, *On Collective Memory* (Lewis A. Coser trans, The University of Chicago Press, 1992).

Haldar, Piyel, 'Forensic Representations of Identity: The *Imago*, the X-Ray and the Evidential Image' (2013) 7(2) *Law and Humanities* 129.

Hale, Sir Matthew, *Historia Placitorum Coronæ: The History of the Pleas of the Crown, Volume 1* (E. and R. Nutt, and R. Gosling, 1736).

Hale, Sir Matthew, *Historia Placitorum Coronæ: The History of the Pleas of the Crown, Volume 2* (E. and R. Nutt, and R. Gosling, 1736).

Hallam, Elizabeth and Jenny Hockey, *Death, Memory and Material Culture* (Berg Publishers, 2010).

Hallam, Elizabeth, Jenny Hockey and Glennys Howarth, *Beyond the Body: Death and Social Identity* (Routledge, 1999).

Hamilton, Carolyn et al. (eds), *Refiguring the Archive* (Springer, 2002).

Harrison, Brian, *Drink and the Victorians: The Temperance Question in England, 1815–1872* (Keele University Press, 1994).

Harrison, Robert Pogue, *Forests: The Shadow of Civilization* (The University of Chicago Press, 1992).

Harrison, Robert Pogue, *The Dominion of the Dead* (The University of Chicago Press, 2003).

Harvard, J.D.J., *The Detection of Secret Homicide: A Study of the Medico-Legal System of Investigation of Sudden and Unexplained Deaths* (Macmillan & Co, 1960).

Harvey, David, *The Urbanization of Capital: Studies in the History and Theory of Capitalist Urbanization* (Johns Hopkins University Press, 1985).

Hawkins, Henry Storry and Henry Giles Shaw, *Manual for Coroners and Magistrates in New South Wales* (W.A. Gullick, 1914).

Heidegger, Martin, 'Building, Dwelling, Thinking', in *Poetry, Language, Thought* (Albert Hofstadter trans, Harper and Row Publishers, 1971).

Hertz, Robert, *Death and the Right Hand* (Rodney Needham and Claudia Needham trans, Cohen & West, [1907 and 1909] 1960 edn).

Hetherington, Kevin, *The Badlands of Modernity: Heterotopia and Social Ordering* (Routledge, 1997).

Hobbes, Thomas, 'Chapter XVI: Of Persons, Authors, and Things Personated', in C. B. Macpherson (ed), *Leviathan* (Penguin Books, 1968).

Hockey, Jenny, *Experience of Death: An Anthropological Account* (Edinburgh University Press, 1990).

Hogg, Russell, 'Law's Other Spaces' (2002) 6 *Law Text Culture* 29.

Howarth, Glennys (ed), *The Changing Face of Death: Historical Accounts of Death and Disposal* (Macmillan Press, 1996).

Howarth, Glennys, *Death and Dying: A Sociological Introduction* (Polity Press, 2007).

Hunnisett, R.F., 'The Origins of the Office of Coroner' (1958) 3 *Royal Historical Society Transactions* 85.

Hunnisett, R.F., *The Medieval Coroner* (Cambridge University Press, 1961).

Hurren, Elizabeth T., *Dissecting the Criminal Corpse: Staging Post-Execution Punishment in Early Modern England* (Palgrave Macmillan, 2016).

Impey, John, *The Office and Duty of Coroners* (J. Butterworth, 1800).

Jacobs, Joseph, 'The Dying of Death' (1899) 72 *Fortnightly Review* 264.

Jalland, Pat, *Death in the Victorian Family* (Oxford University Press, 1996).

Jalland, Pat, *Australian Ways of Death: A Social and Cultural History, 1840–1918* (Oxford University Press, 2002).

Jenkins, Keith, *Re-Thinking History* (Routledge Classics, 2003).

Jervis, Sir John, *A Practical Treatise on the Office and Duties of Coroners* (S. Sweet, 1829).

Jervis, Sir John, *Office and Duties of Coroners* (Sweet & Maxwell Ltd, 7th edn, 1927 [1829]).

Johnston, Alexander James, *A Handy Book for the Coroners of New Zealand* (Government Printer, 1868).

Jones, Pauline (ed), *Historical Records of Victoria – Volume One: Beginnings of Permanent Government* (Melbourne University Press, 1981).

Kafka, Ben, *The Demon of Writing: Powers and Failures of Paperwork* (Zone Books, 2012).

Kassabian, Mihran K., 'The Roentgen Rays in Forensic Medicine' (1901) 19 *Medico-Legal Journal* 407.

Kellehear, Allan, *Death and Dying in Australia* (Oxford University Press, 2000).

Kellehear, Allan, *A Social History of Dying* (Cambridge University Press, 2007).

Kelsen, Hans, *Pure Theory of Law* (University of California Press, 1967).

Kirkby, Diane and Catharine Coleborne (eds), *Law History Colonialism: The Reach of Empire* (Manchester University Press, 2001).

Kirton-Darling, Edward, 'Searching for Pigeons in the Belfry: The Inquest, the Abolition of the Deodand and the Rise of the Family' (2014) 1 *Law, Culture and the Humanities*. http://journals.sagepub.com/doi/full/10.1177/17438721145 60701

Klaver, Elizabeth, 'Introduction', in Elizabeth Klaver (ed), *Images of the Corpse: From the Renaissance to Cyberspace* (University of Wisconsin Press, 2004).

Knight, Bernard, *The Post-Mortem Technician's Handbook: A Manual of Mortuary Practice* (Blackwell Scientific Publications, 1984).

Kostal, Rande W., *Law and English Railway Capitalism, 1825–1875* (Clarendon Press, 1994).

Kübler-Ross, Elizabeth, *On Death and Dying* (Simon and Schuster, 1969).

Laqueur, Thomas W., *The Work of the Dead: A Cultural History of Mortal Remains* (Princeton University Press, 2015).

Latour, Bruno, 'Drawing Things Together', in Michael Lynch and Steve Woolgar (eds), *Representation in Scientific Practice* (MIT Press, 1990).

Latour, Bruno, *The Making of Law: An Ethnography of the Conseil d'Etat* (Marina Brilman and Alain Pottage trans, Polity Press, 2010).

Lawson, Frederick Henry, 'The Creative Use of Legal Concepts' (1957) 32 *New York University Law Review* 907.

Lee, John G., M.D., *Hand-Book for Coroners* (W. Brotherhead, 1881).

Leiboff, Marett, 'Law, Muteness and the Theatrical in Law's Theatrical Presence' (2010) 14 *Law Text Culture* 384.

Leslie, Myles, 'Reforming the Coroner: Death Investigation Manuals in Ontario 1863–1894' (2008) 100(2) *Ontario History* 221.

Leslie, Stanley Myles MacKenzie, *Speaking for the Dead: Coroners, Institutional Structures and Risk Management* (PhD Thesis, Centre for Criminology and Sociolegal Studies, University of Toronto, 2011).

Levine, Harry III, *Medical Imaging* (ABC-CLIO, Santa Barbara, 2010).

Lewis, Milton J., 'Medicine in Colonial Australia, 1788–1900' (2014) 201(1) *Medical Journal of Australia* S5.

Littlejohn, Henry, 'Photography and Criminal Inquiries' (1896) VIII *The Juridical Review: A Journal of Legal and Political Science* 13.

Loar, Carol, 'Medical Knowledge and the Early Modern English Coroner's Inquest' (2010) 23(3) *Social History of Medicine* 475.

Lowther, George, 'A Knowledge of Medicine an Essential Requirement in a Coroner' (1839) 1 *The Lancet* 578.

MacDonagh, Oliver, 'The Nineteenth-Century Revolution in Government: A Reappraisal' (1958) 1(1) *The Historical Journal* 52.

MacDonald, Helen, *Human Remains: Dissection and Its Histories* (Yale University Press, 2006).

MacDonald, Helen, 'The Anatomy Inspector and the Government Corpse' (2009) 6 (2) *History Australia* 401.

MacDonald, Helen, *Possessing the Dead: The Artful Science of Anatomy* (Melbourne University Press, 2010).

MacNevin, Thomas E., *Manual for Coroners and Magistrates in New South Wales: Being a Practical Guide to the Proceedings of the Coroner's Court and to the Holding of Magisterial Inquiries in Lieu of Inquests by Justices of the Peace* (Government Printer, 1st edn, 1875).

MacNevin, Thomas E., *Manual for Coroners and Magistrates in New South Wales: Being a Practical Guide to the Proceedings of the Coroner's Court and to the Holding of Magisterial Inquiries in Lieu of Inquests by Justices of the Peace* (Government Printer, 2nd edn, 1884).

MacNevin, Thomas E., *Manual for Coroners and Magistrates in New South Wales: Being a Practical Guide to the Proceedings of the Coroner's Court and to the Holding of Magisterial Inquiries in Lieu of Inquests by Justices of the Peace* (Charles Potter, Government Printer, 3rd edn, 1895).

Madoff, Ray, *Immortality and the Law: The Rising Power of the American Dead* (Yale University Press, 2010).

Male, George E., *An Epitome of Juridical or Forensic Medicine; for the Use of Medical Men, Coroners, and Barristers* (T. and G. Underwood, 1816).

Manderson, Desmond (ed), *Courting Death: The Law of Mortality* (Pluto Press, 1999).

Marshall, Tim, *Murdering to Dissect: Grave-Robbing, Frankenstein and the Anatomy Literature* (Manchester University Press, 1995).

Mathews, Robert Hamilton, *Handbook to Magisterial Inquiries in New South Wales: Being a Practical Guide for Justices of the Peace in Holding Inquiries Respecting Deaths* (Little and Company, 4th edn, 1902).

Matthews, Daniel, 'From Jurisdiction to Juriswriting: At the Expressive Limits of the Law' (2014) 1 *Law, Culture and the Humanities* 21.

Mauss, Marcel, 'A Category of the Human Mind: The Notion of Person; the Notion of Self' (W.D. Halls trans) in Michael Carrithers, Steven Collins and Steven Lukes (eds), *The Category of the Person: Anthropology, Philosophy, History* (Cambridge University Press, 1985).

Mauss, Marcel, 'Technology (1935/1947)', *in Techniques, Technology and Civilisation* (Dominique Lussier trans, Durkheim Press, 2006).

Mawani, Renisa, 'Law's Archive' (2012) 8 *Annual Review of Law and Social Science* 337.

McKeough, Jill, 'Origins of the Coronial Jurisdiction' (1983) 6 *University of New South Wales Law Journal* 191.

McVeigh, Shaun, 'Law as (More or Less) Itself: On Some Not Very Reflective Elements of Law' (2014) 4 *University of California Irvine Law Review* 471.

Mellor, Philip A. and Chris Shilling, 'Modernity, Self-Identity and the Sequestration of Death' (1993) 27(3) *Sociology* 411.

Minson, Jeffrey, *Bureaucratic Culture and the Management of Sexual Harassment* (Cultural Policy Studies, Occasional Paper No. 12, Institute for Cultural Studies, Division of Humanities, Griffith University, 1991).

Minson, Jeffrey, 'Holding on to Office', in David Burchell and Andrew Leigh (eds), *The Prince's New Clothes: Why Do Australians Dislike Their Politicians?* (UNSW Press, 2002).

Mitchell, Allan, 'The Paris Morgue as a Social Institution in the Nineteenth Century' (1976a) 4 *Francia* 581.

Mitchell, Ann, *Youl, Richard (1821–1897)* (National Centre of Biography, Australian National University, 1976b, Australian Dictionary of Biography). http://adb.anu.edu.au/biography/youl-richard-4900/text8201

Mitford, Jessica, *The American Way of Death* (Simon and Schuster, 1st edn, 1963).

Mnookin, Jennifer L., 'The Image of Truth: Photographic Evidence and the Power of Analogy' (1998) 10(1) *Yale Journal of Law and the Humanities* 1.

Mohr, Richard, 'Flesh and the Person' (2008) 29 *Australian Feminist Law Journal* 31.

Mollison, Crawford Henry, *Lectures on Forensic Medicine* (University of Melbourne, 1921).

Moore, Niamh, Andrea Salter, Liz Stanley and Maria Tamboukou, *The Archive Project: Archival Research in the Social Sciences* (Routledge, 2016).

Mussawir, Edward, *Jurisdiction in Deleuze: The Expression and Representation of Law* (Routledge, 2011).

Mussawir, Edward and Connal Parsley, 'The Law of Persons Today: At the Margins of Jurisprudence' (2017) 11(1) *Law and Humanities* 44.

Naffine, Ngaire, 'Who are Law's Persons? from Cheshire Cats to Responsible Subjects' (2003) 66 *The Modern Law Review* 346.

Neild, James Edward, 'Introductory Lecture to the Course of Forensic Medicine, 12 March 1866' (1866) 5 *Australian Medical Journal* 144.

Neild, James Edward, *Address Delivered to the Annual Meeting of the Victorian Branch of the British Medical Association, July 28, 1882* (Stillwell & Co, 1882).

Nekam, Alexander, *The Personality Conception of the Legal Entity* (Harvard University Press, 1938).

Nettelbeck, Amanda and Robert Foster, 'Colonial Judiciaries, Aboriginal Protection and South Australia's Policy of Punishing "With Exemplary Severity"' (2010) 41(3) *Australian Historical Studies* 319.

Nora, Pierre, 'Between Memory and History: *Les Lieux De Mémoire*' (Marc Roudebush trans, 1989) 26 *Representations* 7.

Osborne, Thomas, 'Bureaucracy as a Vocation: Governmentality and Administration in Nineteenth-Century Britain' (1994) 7(3) *Journal of Historical Sociology* 289.

Osborne, Thomas, 'The Ordinariness of the Archive' (1999) 12(2) *History of the Human Sciences* 51.

Parker, James E.K., *Acoustic Jurisprudence: Listening to the Trial of Simon Bikindi* (Oxford University Press, 2015).

Parsley, Connal, 'The Mask and Agamben: The Transitional Juridical Techniques of Legal Relation' (2010) 14 *Law Text Culture* 12.

Pasveer, Bernike, 'Knowledge of Shadows: The Introduction of X-Ray Images in Medicine' (1989) 11(4) *Sociology of Health and Illness* 360.

Paton, G.A., 'The Development of Forensic Medicine', in John Barry and R. J. Wright-Smith (eds), *The Proceedings of the Medico-Legal Society of Victoria 1939–1940–1941* (Brown, Prior, Anderson, 1941) Vol. 4.

Paxson, James J., *The Poetics of Personification* (Cambridge University Press, 1994).

Philadelphoff-Puren, Nina and Peter Rush, 'Fatal (F)laws: Law, Literature and Writing' (2003) 14 *Law and Critique* 191.

Plueckhahn, Vernon D., *Lectures on Forensic Medicine and Pathology* (University of Melbourne Printing Services, 5th edn, 1982).

Plunkett, John H., *The Australian Magistrate or A Guide to the Duties of A Justice of the Peace for the Colony of New South Wales* (Gazette Office, 1835).

Pollock, Sir Frederick and Frederic William Maitland, *The History of English Law before the Time of Edward I* (Cambridge University Press, 1898).

Pounder, Derrick J., 'Death Investigation in Early Colonial South Australia, 1839–40' (1984) 24(4) *Medicine, Science and the Law* 273.

Prior, Lindsay, *The Social Organisation of Death: Medical Discourse and Social Practices in Belfast* (Macmillan, 1989).

Pugliese, Joseph, '"*Super Visum Corporis*": Visuality, Race, Narrativity and the Body of Forensic Pathology' (2002) 14(2) *Law and Literature* 367.

Quintilian, *Institutes of the Orator* (J. Patsall trans, B. Law and J. Wilkie, 1774) Vol. 2.

Ragon, Michel, *The Space of Death: A Study of Funerary Architecture, Decoration, and Urbanism* (Alan Sheridan trans, University Press of Virginia, 1983).

Rajaram, Prem Kumar, *Ruling the Margins: Colonial Power and Administrative Rule in the past and Present* (Routledge, 2014).

Ranson, David, 'The Role of the Pathologist', in Hugh Selby (ed), *The Aftermath of Death: Coronials* (Federation Press, 1992).

Reed, R. Harvey, 'The X-Ray from a Medico-Legal Standpoint' (1898) 30(18) *Journal of the American Medical Association* 1013.

Reid, Jim, *The Life of Dr William Byam Wilmot M.D. (1805–1874)* (unpublished, Victorian Institute of Forensic Medicine, 2001).

Richardson, Ruth, *Death Dissection and the Destitute* (The University of Chicago Press, 2001).

Ricoeur, Paul, *Living up to Death* (David Pellauer trans, The University of Chicago Press, 2009).

Ridener, John, *From Polders to Postmodernism: A Concise History of Archival Theory* (Litwin Books, 2008).

Riffaterre, Michael, 'Prosopopeia' (1985) 69 *Yale French Studies* 107.

Riles, Annelise (ed), *Documents: Artifacts of Modern Knowledge* (University of Michigan Press, 2006).

Röntgen, Wilhelm Conrad, 'On a New Kind of Rays' (1896) 3(59) *Science* 227.

Rose, Nikolas, 'Expertise and the Government of Conduct' (1994) 14 *Studies of Law, Politics and Society* 359.

Sappol, Michael, *A Traffic of Dead Bodies: Anatomy and Embodied Social Identity in Nineteenth-Century America* (Princeton University Press, 2004).

Sawday, Jonathan, *The Body Emblazoned: Dissection and the Human Body in Renaissance Culture* (Routledge, 1995).

Scholl, R.R., 'Coroner's Inquests', in John Barry and R.J. Wright-Smith (eds), *The Proceedings of the Medico-Legal Society of Victoria 1939-1940-1941* (Brown, Prior, Anderson, 1941) Vol. 4.

Schultz, Bernard, *Art and Anatomy in Renaissance Italy* (UMI Research Press, 1985).

Schwartz, Vanessa R., *Spectacular Realities: Early Mass Culture in Fin-de-Siècle Paris* (University of California Press, 1998).

Scobie, Claire, 'The Return of Bones' (2009) 68(4) *Meanjin.* www.meanjin.com.au/editions/volume-68-number-4-2009/article/the-return-of-the-bones

Scott Bray, Rebecca, *The Eschatology of the Image* (PhD Thesis, University of Melbourne, 2001).

Scott Bray, Rebecca, 'Fugitive Performances of Death and Injury' (2006) 10 *Law Text Culture* 41.

Scott, Charles C., 'X-Ray Pictures as Evidence' (1946) 44(5) *Michigan Law Review* 773.

Scott, Orlando F., 'Rontgenograms and their Chrono-Logic Legal Recognition' (1929) 24 *Illinois Law Review* 674.

Sennett, Richard, *Flesh and Stone: The Body and the City in Western Civilisation* (W.W. Norton & Company, 1996).

Sewell, Richard Clarke, *A Treatise on the Law of Coroner* (O. Richards, 1843).

Sim, Joe and Tony Ward, 'The Magistrate of the Poor? Coroners and Deaths in Custody in Nineteenth-Century England', in Michael Clark and Catherine Crawford (eds), *Legal Medicine in History* (Cambridge University Press, 1994).

Smith, Warren, 'Organizing Death: Remembrance and Re-Collection' (2006) 13(2) *Organization* 225.

Smith, William Ramsay, *A Description of Some Tasmanian Skulls* (Australian Museum, 1900).

Smith, William Ramsay, *A Manual for Coroners: Being a Guide to Coronial Inquiries and Inquests in South Australia and Throughout Australasia and in England* (Hussey & Gillingham, 1904).

Smith, William Ramsay, 'The Aborigines of Australians', in *Official Year Book of the Commonwealth of Australia* (Commonwealth Bureau of Census and Statistics, 1909).

Smith, William Ramsay, *Medical Jurisprudence from the Judicial Standpoint* (Stevens and Sons, 1913).

Smith, William Ramsay, *In Southern Seas: Wanderings of a Naturalist* (John Murray, 1924).

Smith, William Ramsay, *Myths and Legends of the Australian Aborigines* (Dover Publications, [1932] 2003 edn).

Smith, William Ramsay and Edward Charles Stirling, *Australian Aborigines: Burial Grounds* (Thomas Gill, 1907–1911).

Smyth, William Henry, *Technocracy Part III: Ways and Means to Gain Industrial Democracy* (Gazette, 1920).

Snyder, Joel, 'Res Ipsa Loquitur', in Lorraine Daston (ed), *Things that Talk: Object Lessons from Art and Science* (Zone Books, 2008).

Sohn, Heidi, 'Heterotopia: Anamnesis of a Medical Term', in Michiel Dehaene and Lieven De Cauter (eds), *Heterotopia and the City: Public Space in a Postcivil Society* (Routledge, 2008).

Steedman, Carolyn, *Dust: The Archive and Cultural History* (Rutgers University Press, 2002).

Stepputat, Finn (ed), *Governing the Dead: Sovereignty and the Politics of Dead Bodies* (Manchester University Press, 2014).

Stoler, Laura Ann, *Along the Archival Grain: Epistemic Anxieties and Colonial Common Sense* (Princeton University Press, 2010).

Strauss, Jonathan, *Human Remains: Medicine, Death, and Desire in Nineteenth-Century Paris* (Fordham University Press, 2012).

Sutton, Teresa, 'The Deodand and Responsibility for Death' (1997) 18(3) *The Journal of Legal History* 44.

Sutton, Teresa, 'The Nature of the Early Law of Deodand' (1999) 30(9) *Cambrian Law Review* 14.

Thorne, Samuel Edmund, *Essays in Legal History* (Hambledon Press, 1985).

Tomlins, P.S., *The Coroner's Guide: A Summary of the Duties, Powers and Liabilities of Coroners from the Most Approved Authorities* (Van Diemen's Land, 1837).

Trabsky, Marc, 'Institutionalising the Public Abattoir in Nineteenth Century Colonial Society' (2014) 40(2) *Australian Feminist Law Journal* 169.

Trabsky, Marc, 'The Custodian of Memories: Coronial Architecture in Nineteenth Century Melbourne' (2015) 24(2) *Griffith Law Review* 199.

Tuitt, Patricia, 'Legal Practice and Modes of Dying: Bruno Latour, Technology and the Critical Legal Instance' (2005) 16 *Law and Critique* 113.

Turnbull, Paul and Michael Pickering (eds), *The Long Way Home: The Meaning and Values of Repatriation* (Berghaghn Books, 2010).

Umfreville, Edward, *Lex Coronatoria: Or, the Office and Duty of Coroners: In Three Parts – Wherein the Theory of the Office Is Distinctly Laid Down; and the Practice Illustrated* (R. Griffiths and T. Becket, 1761).

Valverde, Mariana, *Chronotopes of Law: Jurisdiction, Scale, and Governance* (Routledge, 2015).

van Dijck, José, *The Transparent Body: A Cultural Analysis of Medical Imaging* (University of Washington Press, 2005).

Veitch, Scott, *Law and Irresponsibility: On the Legitimation of Human Suffering* (Routledge-Cavendish, 2007).

Vernant, Jean-Pierre and Françoise Frontisi-Ducroux, 'Features of the Mask in Ancient Greece', in Jean-Pierre Vernant and Pierre Vidal-Naquet (eds), *Myth and Tragedy in Ancient Greece* (Janet Lloyd trans, Zone Books, 1988).

Veyne, Paul, 'Foucault Revolutionizes History', in Arnold I. Davidson (ed), *Foucault and His Interlocutors* (Catherine Porter trans, The University of Chicago Press, 1997).

Vickers, Robert H., *The Powers and Duties of Police Officers and Coroners* (T.H. Flood, 1889).

Vico, Giambattista, *New Science* (David Marsh trans, Penguin Books, 3rd edn, 1999).

Vismann, Cornelia, *Files: Law and Media Technology* (Geoffrey Winthrop-Young trans, Stanford University Press, 2008).

Waldo, F.J., 'The Ancient Office of Coroner' (1910–1911) 8 *Transactions of the Medico-Legal Society* 101.

Wallerstein, Immanuel, *The Modern-World System IV: Centrist Liberalism Triumphant 1789–1914* (University of California Press, 2011).

Walter, Tony, *The Revival of Death* (Routledge, 1994).

Weber, Max, *Economy and Society: An Outline of Interpretive Sociology* (Guenther Roth and Claus Wittich trans, Bedminster Press, 1968).

Weber, Max, *The Protestant Ethic and the Spirit of Capitalism* (Talcott Parsons trans, Routledge, 1992).

Weber, Max, *The Vocation Lectures* (Rodney Livingstone trans, Hackett Publishing Company, 2004).

Wellington, Richard Henslowe, *The King's Coroner* (William Clowes & Sons, 1905).

Whaley, Joachim (ed), *Mirrors of Mortality: Studies in the Social History of Death* (Routledge, 1981).

Wilmot, William B., 'On the Principles of Pathology' (1856) 1(1) *Australian Medical Journal* 1.

Wohl, Victoria, *Law's Cosmos: Juridical Discourse in Athenian Forensic Oratory* (Cambridge University Press, 2010).

Wolff, Martin, 'On the Nature of Legal Persons' (1938) 54 *The Law Quarterly Review* 494.

Index

administration: bureaucratic *see* bureaucracy; compared to patronage 80; of coronial law and procedure *see* coronial law, administration of; and the examination 94–98; of justice 28, 78, 90; rationalisation 80, 85

affirmations 69

Agamben, Giorgio 13n44, 59n74, 76n48

amercements 3, 24, 45, 90

anatomy 15, 40–46; unlawful dissection 72–76

anthologisation, activity of 68–69, 74, 84

apparatus (*dispositif*) 6, 59

'archival thinking' 9–10

archives 9–10, 85, 94–102; *see also* cases; files

archivist, persona of 90, 102

Ariès, Philippe 11, 13, 19

asylums, 52n48 72, 74–75, 82

Australian colonies 6–7, 47, 50–58, 65

autopsy *see* post-mortem examination

Bataille, Georges 4n12, 13n44

Benjamin, Walter 11

biography 15–16, 85–86, 96–102

biopower 5

The Birth of the Clinic (Foucault) 44–45, 58, 60, 108

black box 45

book-keeping 4, 69; *see also* record-keeping

Boys, William F.A. 66, 78–79

British Empire 6–7, 23, 25, 29

building 15, 18, 27, 28–30, 36, 37

bureaucracy: and irresponsibility 83; and justice 81; and office 80–82; rationality 12, 80–82; and vocation 82; *see also* administration, bureaucratic

burial practices 19, 21–23, 34–35, 46, 73; *see also* cemeteries

Burney, Ian 26n30, 27–28, 30, 47, 55–57

cadavers 24, 39–40, 42–43; *see also* dead bodies

Candler, Samuel Curtis 35, 86–89, 102

capitalism 4, 80

care: for the dead *see* the dead, caring for; as a duty of the state 5, 8; ethics of 71, 83, 99, 102, 112 (*see also* office, ethics of); as maintaining legal relations with the dead 8, 99

Carter, Paul 21, 29

cases 95–100; *see also* archives; documents; files

cemeteries 19, 20–21, 36, 72; *see also* burial practices; necropolis

de Certeau, Michel 32, 38, 93, 101

character 24, 58, 71, 77–79, 82, 84; *see also* persona

Cicero 60, 76

circulars 64, 66–71, 74

'the city of the living' 21, 32

civil service 71, 77, 79–80, 85, 88

clinical experience 44, 58

colonisation 6–7, 22–23, 29, 36

common law 1, 34–35, 43n15, 69, 74, 83, 87–89, 92, 94, 100

contempt of court 86–89, 91, 94

coroners: accountability of 74, 80, 83, 84, 99; appointment of 21–23, 36, 48, 53, 77–80; calls for abolition of 52–54, 57; conflict with medical practitioners 50–54; dereliction of duty 17, 18, 23; discretion of 75–76, 78, 82; duties of 3, 16, 22, 29, 45–49, 53, 54, 58, 65–79, 84, 94, 98, 99; fiscal origins of 2–4, 24, 90; itinerant